FOOD INSECURITY
AND DISEASE

Prevalence, Policy, and Politics

FOOD INSECURITY AND DISEASE

Prevalence, Policy, and Politics

Edited by
Areej Hassan, MD, MPH

Apple Academic Press Inc.	Apple Academic Press Inc.
3333 Mistwell Crescent	9 Spinnaker Way
Oakville, ON L6L 0A2	Waretown, NJ 08758
Canada	USA

©2017 by Apple Academic Press, Inc.

First issued in paperback 2021

Exclusive worldwide distribution by CRC Press, a member of Taylor & Francis Group

No claim to original U.S. Government works

ISBN 13: 978-1-77-463688-6 (pbk)
ISBN 13: 978-1-77-188491-4 (hbk)

Library and Archives Canada Cataloguing in Publication

Food insecurity and disease : prevalence, policy, and politics / edited by Areej Hassan, MD, MPH.

Includes bibliographical references and index.
Issued in print and electronic formats.
ISBN 978-1-77188-491-4 (hardcover).--ISBN 978-1-315-36576-3

1. Food security. 2. Food security--Health aspects.
3. Public health. 4. Nutrition. 5. Food supply. I. Hassan, Areej, editor

| HD9000.5.F648 2017 | 338.1'9 | C2016-907673-3 | C2016-907674-1 |

CIP data on file with US Library of Congress

Apple Academic Press also publishes its books in a variety of electronic formats. Some content that appears in print may not be available in electronic format. For information about Apple Academic Press products, visit our website at **www.appleacademicpress.com** and the CRC Press website at **www.crcpress.com**

About the Editor

Areej Hassan, MD, MPH

Areej Hassan, MD, MPH, is an attending in the Division of Adolescent/Young Adult Medicine at Boston Children's Hospital and Assistant Professor of Pediatrics at Harvard Medical School. She completed her residency training in Pediatrics at Brown University before her fellowship at BCH. In addition to primary care, Dr. Hassan focuses her clinical interests on reproductive endocrinology and global health. She also maintains an active role in medical education and has particular interest in building and developing innovative teaching tools through open educational resources. She currently teaches, consults, and is involved in pediatric and adolescent curricula development at multiple sites abroad in Central America and Southeast Asia.

Contents

List of Contributors..*ix*

Acknowledgments and How to Cite...*xiii*

Introduction ....*xv*

Part I: Defining Food Security and Insecurity.. 1

1. Definitions of Food Security 3
United States Department of Agriculture

Part II: Food Insecurity and Mental Health... 9

**2. Food Insecurity in Adults with Mood Disorders: Prevalence
    Estimates and Associations with Nutritional and Psychological Health 11**
Karen M. Davidson and Bonnie J. Kaplan

**3. Household Food Insecurity and Mental Distress Among Pregnant
    Women in Southwestern Ethiopia: A Cross Sectional Study Design 27**
Mulusew G. Jebena, Mohammed Taha, Motohiro Nakajima, Andrine Lemieux,
Fikre Lemessa, Richard Hoffman, Markos Tesfaye, Tefera Belachew,
Netsanet Workineh, Esayas Kebede, Teklu Gemechu, Yinebeb Tariku,
Hailemariam Segni, Patrick Kolsteren, and Mustafa al'Absi

Part III: Food Insecurity and HIV 45

**4. Is Food Insecurity Associated with HIV Risk? Cross-Sectional
    Evidence from Sexually Active Women in Brazil 47**
Alexander C. Tsai, Kristin J. Hung, and Sheri D. Weiser

**5. Food Insecurity Is a Barrier to Prevention of Mother-to-Child HIV
    Transmission Services in Zimbabwe: A Cross-Sectional Study 69**
Sandra I. McCoy, Raluca Buzdugan, Angela Mushavi, Agnes Mahomva,
Frances M. Cowan, and Nancy S. Padian

**6. A Pre-Post Pilot Study of Peer Nutritional Counseling and
    Food Insecurity and Nutritional Outcomes among Antiretroviral
    Therapy Patients in Honduras.. 89**
Kathryn P. Derose, Melissa Felician, Bing Han, Kartika Palar, Blanca Ramírez,
Hugo Farías, and Homero Martínez

7.   Relationship between Food Insecurity and Mortality among
     HIV-Positive Injection Drug Users Receiving Antiretroviral
     Therapy in British Columbia, Canada..105

     Aranka Anema, Keith Chan, Yalin Chen, Sheri Weiser, Julio S. G. Montaner,
     and Robert S. Hogg

8.   *Shamba Maisha:* **Pilot Agricultural Intervention for Food Security**
     and HIV Health Outcomes in Kenya: Design, Methods, Baseline
     Results and Process Evaluation of a Cluster-Randomized
     Controlled Trial...125

     Craig R. Cohen, Rachel L. Steinfeld, Elly Weke, Elizabeth A. Bukusi,
     Abigail M. Hatcher, Stephen Shiboski, Richard Rheingans, Kate M. Scow,
     Lisa M. Butler, Phelgona Otieno, Shari L. Dworkin, and Sheri D. Weiser

Part IV: Food Security and Obesity and Diabetes 159

9.   Challenges of Diabetes Self-Management in Adults Affected by
     Food Insecurity in a Large Urban Centre of Ontario, Canada 161

     Justine Chan, Margaret DeMelo, Jacqui Gingras, and Enza Gucciardi

10. Children's Very Low Food Security is Associated with Increased
     Dietary Intakes in Energy, Fat, and Added Sugar among
     Mexican-Origin Children (6-11 Y) in Texas Border *Colonias*....................179

     Joseph R. Sharkey, Courtney Nalty, Cassandra M. Johnson, and Wesley R. Dean

11. Obesity Prevention and National Food Security: A Food
     Systems Approach 199

     Lila Finney Rutten, Amy Lazarus Yaroch, Heather Patrick, and Mary Story

Part V: Policy, Power, and Politics 219

12. Food Sovereignty: Power, Gender, and the Right to Food 221

     Rajeev C. Patel

13. Big Food, Food Systems, and Global Health...231

     David Stuckler and Marion Nestle

Keywords 241

Author Notes 243

Index 251

List of Contributors

Mustafa al'Absi
Duluth Medical Research Institute, Department of Bio behavioral Health and Population Sciences, University of Minnesota Medical School

Aranka Anema
British Columbia Centre for Excellence in HIV/AIDS, St. Paul's Hospital, Vancouver, British Columbia, Canada, Department of Medicine, Faculty of Medicine, University of British Columbia, Vancouver, British Columbia, Canada

Tefera Belachew
Population and Family Health, Jimma University

Elizabeth A. Bukusi
Centre for Microbiology Research, Kenya Medical Research Institute

Lisa M. Butler
Boston Children's Hospital and Harvard Medical School

Raluca Buzdugan
University of California

Justine Chan
Ryerson University, 350 Victoria Street, Toronto, ON, Canada M5B 2K3

Keith Chan
Department of Medicine, Faculty of Medicine, University of British Columbia, Vancouver, British Columbia, Canada

Yalin Chen
Department of Medicine, Faculty of Medicine, University of British Columbia, Vancouver, British Columbia, Canada

Craig R. Cohen
Department of Obstetrics, Gynecology & Reproductive Sciences, University of California San Francisco; Center of Expertise in Women's Health & Empowerment, University of California Global Health Institute

Frances M. Cowan
Centre for Sexual Health and HIV Research; University College London

Karen M. Davison
Department of Community Health Sciences, University of Calgary
Department of Biology, Health Science Program, Kwantlen Polytechnic University

Wesley R. Dean
Program for Research in Nutrition and Health Disparities, School of Rural Public Health, Texas A&M Health Science Center, MS 1266

Margaret DeMelo
University Health Network, 399 Bathurst Street, Toronto, ON, Canada M5T 2S

Kathryn P. Derose
Health Program, RAND Corporation

Shari L. Dworkin
Departments of Social and Behavioral Sciences, University of California San Francisco; Center of Expertise in Women's Health & Empowerment, University of California Global Health Institute

Hugo Farías
Regional Office for Latin America and the Caribbean, United Nations World Food Program

Melissa Felician
Pardee RAND Graduate School

Teklu Gemechu
Department of Psychology, Jimma University

Jacqui Gingras
Ryerson University, 350 Victoria Street, Toronto, ON, Canada M5B 2K3

Enza Gucciardi
Ryerson University, 350 Victoria Street, Toronto, ON, Canada M5B 2K3

Bing Han
Health Program, RAND Corporation

Abigail M. Hatcher
Department of Obstetrics, Gynecology & Reproductive Sciences, University of California San Francisco; Wits Reproductive Health and HIV Institute, University of the Witwatersrand

Richard Hoffman
Duluth Medical Research Institute, Department of Bio behavioral Health and Population Sciences, University of Minnesota Medical School

Robert S. Hogg
British Columbia Centre for Excellence in HIV/AIDS, St. Paul's Hospital, Vancouver, British Columbia, Canada, Faculty of Health Sciences, University of British Columbia, Burnaby, British Columbia, Canada

Kristin J. Hung
Department of Obstetrics and Gynecology, Beth Israel Deaconess Medical Center, Boston, Massachusetts, United States of America

Mulusew G. Jebena
Population and Family Health, Jimma University; Department of Food Safety and Food Quality, Ghent University, CoupureLinks

Cassandra M. Johnson
UNC Center for Health Promotion and Disease Prevention and Department of Nutrition, UNC Gillings School of Global Public Health

Bonnie J. Kaplan
Department of Community Health Sciences, University of Calgary
Department of Pediatrics, University of Calgary
The Alberta Children's Hospital Research Institute

Esayas Kebede
Department of Internal Medicine, Jimma University

Patrick Kolsteren
Department of Food Safety and Food Quality, Ghent University, CoupureLinks

Fikre Lemessa
Department of Horticulture and Plant Sciences, Jimma University

Andrine Lemieux
Duluth Medical Research Institute, Department of Bio behavioral Health and Population Sciences, University of Minnesota Medical School

Agnes Mahomva
Elizabeth Glaser Pediatric AIDS Foundation

Homero Martínez
Health Program, RAND Corporation; Hospital Infantil de México Federico Gómez

Sandra I. McCoy
University of California

Julio S. G. Montaner
British Columbia Centre for Excellence in HIV/AIDS, St. Paul's Hospital, Vancouver, British Columbia, Canada, Department of Medicine, Faculty of Medicine, University of British Columbia, Vancouver, British Columbia, Canada

Angela Mushavi
Ministry of Health and Child Welfare

Motohiro Nakajima
Duluth Medical Research Institute, Department of Bio behavioral Health and Population Sciences, University of Minnesota Medical School

Courtney Nalty
Program for Research in Nutrition and Health Disparities, School of Rural Public Health, Texas A&M Health Science Center, MS 1266

Marion Nestle
Department of Nutrition, Food Studies, and Public Health, New York University, New York, New York, United States of America, Department of Nutritional Sciences, Cornell University, Ithaca, New York, United States of America

Phelgona Otieno
Centre for Clinical Research, Kenya Medical Research Institute

Nancy S. Padian
University of California

Kartika Palar
Division of HIV, ID and Global Medicine, Department of Medicine, University of California San Francisco

Rajeev C. Patel
School of Development Studies, University of KwaZulu-Natal, Durban, KwaZulu-Natal, South Africa

Heather Patrick
Health Behaviors Research Branch, National Cancer Institute, 6130 Executive Boulevard, MSC 7335, Bethesda, MD 20892, USA

Blanca Ramírez
Honduran Country Office, United Nations World Food Program

Richard Rheingans
Department of Environmental and Global Health, University of Florida

Lila Finney Rutten
Division of Epidemiology, Department of Health Sciences Research, Mayo Clinic, 200 First Street SW, Rochester, MN 55905, USA; Division of Epidemiology and Community Health, School of Public Health, University of Minnesota, 1300 S 2nd Street, Suite 300, Minneapolis, MN 55454, USA

Kate M. Scow
Department of Soil Science and Soil Microbial Biology, University of California Davis

Hailemariam Segni
Department of Obstetrics and Gynecology, Jimma University

Joseph R. Sharkey
Program for Research in Nutrition and Health Disparities, School of Rural Public Health, Texas A&M Health Science Center, MS 1266

Stephen Shiboski
Departments of Epidemiology and Biostatistics, University of California San Francisco

Rachel L. Steinfeld
Department of Obstetrics, Gynecology & Reproductive Sciences, University of California San Francisco

Mary Story
Division of Epidemiology, Department of Health Sciences Research, Mayo Clinic, 200 First Street SW, Rochester, MN 55905, USA

David Stuckler
Department of Sociology, University of Cambridge, Cambridge, United Kingdom, Department of Public Health & Policy, London School of Hygiene & Tropical Medicine, London, United Kingdom

Mohammed Taha
Department of Epidemiology, Jimma University

Yinebeb Tariku
Department of Chemistry, Jimma University

Markos Tesfaye
Department of Psychiatry, Jimma University

Alexander C. Tsai
Robert Wood Johnson Health and Society Scholars Program, Harvard University, Cambridge, Massachusetts, United States of America, Center for Global Health, Massachusetts General Hospital, Boston, Massachusetts, United States of America

Sheri D. Weiser
Division of HIV/AIDS, San Francisco General Hospital, University of California San Francisco, San Francisco, California, United States of America; Center for AIDS Prevention Studies, University of California San Francisco, San Francisco, California, United States of America

Elly Weke
Centre for Microbiology Research, Kenya Medical Research Institute

Netsanet Workineh
Department of Pediatrics and Child Health, Jimma University

Amy Lazarus Yaroch
Gretchen Swanson Center for Nutrition, 505 Durham Research Plaza, Omaha, NE 68105, USA

Acknowledgments and How to Cite

The editor and publisher thank each of the authors who contributed to this book. Many of the chapters in this book were previously published elsewhere. To cite the work contained in this book and to view the individual permissions, please refer to the citation at the beginning of each chapter. The editor carefully selected each chapter individually to provide a nuanced look at food insecurity and its connection to disease.

Introduction

We know that food insecurity and disease are inextricably linked. The articles selected for this compendium reinforce that message by specifically linking food insecurity to various forms of chronic disease, including HIV/AIDS and obesity, as well as mental health issues. The research in the fifth section of this book then goes a step further by asking, "What next?" In other words, how can we shape politics and policy to address this urgent international crisis?

The quality of the research gathered here is incredibly high. The authors have done much to advance our understanding of this issue—and they have provided us with a solid foundation on which to build well-informed clinical practice, further research, and effective future policy.

—*Areej Hassan, MD*

According to the USDA, food security means access by all people at all times to enough food for an active, healthy life. We begin our compendium with the USDA's definitions of food security and insecurity.

Because little is known about food insecurity in people with mental health conditions, the authors of chapter 2 investigated relationships among food insecurity, nutrient intakes, and psychological functioning in adults with mood disorders. Data from a study of adults randomly selected from the membership list of the Mood Disorder Association of British Columbia (n = 97), Canada, were analyzed. Food insecurity status was based on validated screening questions asking if in the past 12 months did the participant, due to a lack of money, worry about or not have enough food to eat. Nutrient intakes were derived from 3-day food records and compared to the Dietary Reference Intakes (DRIs). Psychological functioning measures included Global Assessment of Functioning, Hamilton Depression scale, and Young Mania Rating Scale. Using binomial tests of two proportions, Mann–Whitney U tests, and Poisson regression authors examined: (1) food insecurity prevalence between the study respondents and a general population sample from the British Columbia Nutrition Survey (BCNS; n = 1,823); (2) differences in nutrient intakes based on food insecurity status; and (3) associations of food insecurity and psychological functioning using bivariate and Poisson regression statistics. In comparison to the general population (BCNS), food insecurity was significantly more prevalent in the adults with mood disorders (7.3% in BCNS vs 36.1%; $p < 0.001$). Respondents who were food-insecure had lower median intakes of carbohydrates and vitamin C ($p < 0.05$). In addition, a higher proportion of those reporting food insecurity had protein, folate, and zinc intakes below the DRI benchmark of potential inadequacy ($p < 0.05$). There was significant association between food insecurity and mania symptoms (adjusted prevalence ratio = 2.37, 95% CI 1.49–3.75, $p < 0.05$). The authors found that food insecurity is associated with both nutritional and psychological health in adults with mood disorders. Investigation of interventions aimed at food security and income can help establish its role in enhancing mental health.

There are compelling theoretical and empirical reasons that link household food insecurity to mental distress in the setting where both problems are common. However, little is known about their association during pregnancy in Ethiopia. A cross-sectional study is reported in chapter 3 that was conducted to examine the association of household food insecurity with mental distress during pregnancy. Six hundred and forty-two pregnant women were recruited from 11 health centers and one hospital. Probability proportional to size (PPS) and consecutive sampling techniques were employed to recruit study subjects until the desired sample size was obtained. The Self Reporting Questionnaire (SRQ-20) was used to measure mental distress and a 9-item Household Food Insecurity Access Scale was used to measure food security status. Descriptive and inferential statistics were computed accordingly. Multivariate logistic regression was used to estimate the effect of food insecurity on mental distress. Fifty-eight of the respondents (9 %) were moderately food insecure and 144 of the respondents (22.4 %) had mental distress. Food insecurity was also associated with mental distress. Pregnant women living in food insecure households were 4 times more likely to have mental distress than their counterparts (COR = 3.77, 95% CI: 2.17, 6.55). After controlling for confounders, a multivariate logistic regression model supported a link between food insecurity and mental distress (AOR = 4.15, 95% CI: 1.67, 10.32). The study found a significant association between food insecurity and mental distress. However, the mechanism by which food insecurity is associated with mental distress is not clear. Further investigation is therefore needed to understand either how food insecurity during pregnancy leads to mental distress or weather mental distress is a contributing factor in the development of food insecurity.

Understanding how food insecurity among women gives rise to differential patterning in HIV risks is critical for policy and programming in resource-limited settings. This is particularly the case in Brazil, which has undergone successive changes in the gender and socio-geographic composition of its complex epidemic over the past three decades. The authors of chapter 4 used data from a national survey of Brazilian women to estimate the relationship between food insecurity and HIV risk. They used data on 12,684 sexually active women from a national survey conducted in Brazil in 2006–2007. Self-reported outcomes were (a) consistent condom use, defined as using a condom at each occasion of sexual intercourse in the previous 12 mo; (b) recent condom use, less stringently defined as using a condom with the most recent sexual partner; and (c) itchy vaginal discharge in the previous 30 d, possibly indicating presence of a sexually transmitted infection. The primary explanatory variable of interest was food insecurity, measured using the culturally adapted and validated Escala Brasiliera de Segurança Alimentar. In multivariable logistic regression models, severe food insecurity with hunger was associated with a reduced odds of consistent condom use in the past 12 mo (adjusted odds ratio [AOR]=0.67; 95% CI, 0.48–0.92) and condom use at last sexual intercourse (AOR=0.75; 95% CI, 0.57–0.98). Self-reported itchy vaginal discharge was associated with all categories of food insecurity (with AORs ranging from 1.46 to 1.94). In absolute terms, the effect sizes were large in magnitude across all outcomes. Underweight and/or lack of control in sexual relations did not appear to mediate the observed associations. Severe food insecurity with hunger was associated with reduced odds of condom use and increased odds of itchy vaginal discharge, which is potentially indicative of sexually transmitted infection, among sexually active women in Brazil. Interventions targeting food insecurity may have beneficial implications for HIV prevention in resource-limited settings.

Food insecurity (FI) is the lack of physical, social, and economic access to sufficient food for dietary needs and food preferences. The authors of chapter 5 examined the association between FI and women's uptake of services to prevent mother-to-child HIV transmission (MTCT) in Zimbabwe. They analyzed cross-sectional data collected in 2012 from women living in five of ten provinces. Eligible women were ≥16 years old, biological mothers of infants born 9–18 months before the interview, and were randomly selected using multi-stage cluster sampling. Women and infants were tested for HIV and interviewed about health service utilization during pregnancy, delivery, and post-partum. The authors assessed FI in the past four weeks using a subset of questions from the Household Food Insecurity Access Scale and classified women as living in food secure, moderately food insecure, or severely food insecure households. The weighted population included 8,790 women. Completion of all key steps in the PMTCT cascade was reported by 49%, 45%, and 38% of women in food secure, moderately food insecure, and severely food insecure households, respectively (adjusted prevalence ratio (PRa) = 0.95, 95% confidence interval (CI): 0.90, 1.00 (moderate FI vs. food secure), PRa = 0.86, 95% CI: 0.79, 0.94 (severe FI vs. food secure)). Food insecurity was not associated with maternal or infant receipt of ART/ARV prophylaxis. However, in the unadjusted analysis, among HIV-exposed infants, 13.3% of those born to women who reported severe household food insecurity were HIV-infected compared to 8.2% of infants whose mothers reported food secure households (PR = 1.62, 95% CI: 1.04, 2.52). After adjustment for covariates, this association was attenuated (PRa = 1.42, 95% CI: 0.89, 2.26). There was no association between moderate food insecurity and MTCT in unadjusted or adjusted analyses (PRa = 0.68, 95% CI: 0.43, 1.08). Among women with a recent birth, food insecurity is inversely associated with service utilization in the PMTCT cascade and severe household food insecurity may be positively associated with MTCT. These preliminary findings support the assessment of FI in antenatal care and integrated food and nutrition programs for pregnant women to improve maternal and child health.

Food insecurity and poor nutrition are key barriers to anti-retroviral therapy (ART) adherence. Culturally-appropriate and sustainable interventions that provide nutrition counseling for people on ART and of diverse nutritional statuses are needed, particularly given rising rates of overweight and obesity among people living with HIV (PLHIV). As part of scale-up of a nutritional counseling intervention, the authors of chapter 6 recruited and trained 17 peer counselors from 14 government-run HIV clinics in Honduras to deliver nutritional counseling to ART patients using a highly interactive curriculum that was developed after extensive formative research on locally available foods and dietary patterns among PLHIV. All participants received the intervention; at baseline and 2 month follow-up, assessments included: 1) interviewer-administered, in-person surveys to collect data on household food insecurity (15-item scale), nutritional knowledge (13-item scale), dietary intake and diversity (number of meals and type and number of food groups consumed in past 24 h); and 2) anthropometric measures (body mass index or BMI, mid-upper arm and waist circumferences). The authors used multivariable linear regression analysis to examine changes pre-post in food insecurity and the various nutritional outcomes while controlling for baseline characteristics and clinic-level clustering. Of 482 participants at baseline, the authors had complete follow-up data on 356 (74%), of which 62% were women, median age was 39, 34% reported having paid work, 52% had completed primary school, and 34% were overweight or obese. In multivariate analyses adjusting for gender, age, household size, work status, and education, the authors found that between baseline and follow-up, household food insecurity decreased significantly among all participants ($\beta = -0.47$, $p < .05$) and among those with children under 18 ($\beta = -1.16$, $p < .01$), while nutritional knowledge and dietary intake and diversity also significantly improved, ($\beta = 0.88$, $p < .001$; $\beta = 0.30$, $p < .001$; and $\beta = 0.15$, $p < .001$, respectively). Nutritional status (BMI, mid-arm and waist circumferences) showed no significant changes, but the brief follow-up period may not have been sufficient to detect changes. A peer-delivered nutritional counseling intervention for PLHIV was associated with improvements in dietary quality and reduced food insecurity among a population of diverse nutritional statuses. Future research should examine if such an intervention can improve adherence among people on ART.

Little is known about the potential impact of food insecurity on mortality among people living with HIV/AIDS. The authors of chapter 7 examined the potential relationship between food insecurity and all-cause mortality among HIV-positive injection drug users (IDU) initiating antiretroviral therapy (ART) across British Columbia (BC). Cross-sectional measurement of food security status was taken at participant ART initiation. Participants were prospectively followed from June 1998 to September 2011 within the fully subsidized ART program. Cox proportional hazard models were used to ascertain the association between food insecurity and mortality, controlling for potential confounders. Among 254 IDU, 181 (71.3%) were food insecure and 108 (42.5%) were hungry. After 13.3 years of median follow-up, 105 (41.3%) participants died. In multivariate analyses, food insecurity remained significantly associated with mortality (adjusted hazard ratio [AHR] = 1.95, 95% CI: 1.07–3.53), after adjusting for potential confounders. HIV-positive IDU reporting food insecurity were almost twice as likely to die, compared to food secure IDU. Further research is required to understand how and why food insecurity is associated with excess mortality in this population. Public health organizations should evaluate the possible role of food supplementation and socio-structural supports for IDU within harm reduction and HIV treatment programs.

Despite advances in treatment of people living with HIV, morbidity and mortality remains unacceptably high in sub-Saharan Africa, largely due to parallel epidemics of poverty and food insecurity. The authors of chapter 8 conducted a pilot cluster randomized controlled trial (RCT) of a multisectoral agricultural and microfinance intervention (entitled *Shamba Maisha*) designed to improve food security, household wealth, HIV clinical outcomes and women's empowerment. The intervention was carried out at two HIV clinics in Kenya, one randomized to the intervention arm and one to the control arm. HIV-infected

patients >18 years, on antiretroviral therapy, with moderate/severe food insecurity and/or body mass index (BMI) <18.5, and access to land and surface
water were eligible for enrollment. The intervention included: 1) a microfinance loan (~$150) to purchase the farming commodities, 2) a micro-irrigation
pump, seeds, and fertilizer, and 3) trainings in sustainable agricultural practices
and financial literacy. Enrollment of 140 participants took four months, and the
screening-to-enrollment ratio was similar between arms. The authors followed
participants for 12 months and conducted structured questionnaires. They also
conducted a process evaluation with participants and stakeholders 3–5 months
after study start and at study end. Baseline results revealed that participants at
the two sites were similar in age, gender and marital status. A greater proportion
of participants at the intervention site had a low BMI in comparison to participants at the control site (18% vs. 7%, p = 0.054). While median CD4 count was
similar between arms, a greater proportion of participants enrolled at the intervention arm had a detectable HIV viral load compared with control participants
(49% vs. 28%, respectively, p < 0.010). Process evaluation findings suggested
that *Shamba Maisha* had high acceptability in recruitment, delivered strong agricultural and financial training, and led to labor saving due to use of the water
pump. Implementation challenges included participant concerns about repaying loans, agricultural challenges due to weather patterns, and a challenging partnership with the microfinance institution. The authors expect the results from
this pilot study to provide useful data on the impacts of livelihood interventions
and will help in the design of a definitive cluster RCT.

　　The authors of chapter 9 aim to explore how food insecurity affects individuals' ability to manage their diabetes, as narrated by participants living in a large,
culturally diverse urban centre. To this end, the authors underwent a qualitative
study comprising of in-depth interviews, using a semistructured interview guide.
Participants were recruited from the local community, three community health
centres, and a community-based diabetes education centre servicing a low-
income population in Toronto, Ontario, Canada. Twenty-one English-speaking
adults with a diagnosis of diabetes and having experienced food insecurity in the
past year (based on three screening questions) participated in the study. Using
six phases of analysis, the authors used qualitative, deductive thematic analysis
to transcribe, code, and analyze participant interviews. Three themes emerged
from the authors' analysis of participants' experiences of living with food insecurity and diabetes: (1) barriers to accessing and preparing food, (2) social
isolation, and (3) enhancing agency and resilience. Food insecurity appears to
negatively impact diabetes self-management. Healthcare professionals need to

be cognizant of resources, skills, and supports appropriate for people with diabetes affected by food insecurity. Study findings suggest foci for enhancing diabetes self-management support.

Food insecurity among Mexican-origin and Hispanic households is a critical nutritional health issue of national importance. At the same time, nutrition-related health conditions, such as obesity and type 2 diabetes, are increasing in Mexican-origin youth. Risk factors for obesity and type 2 diabetes are more common in Mexican-origin children and include increased intakes of energy-dense and nutrient-poor foods. Chapter 10 assessed the relationship between children's experience of food insecurity and nutrient intake from food and beverages among Mexican-origin children (age 6-11 y) who resided in Texas border colonias. Baseline data from 50 Mexican-origin children were collected in the home by trained promotora-researchers. All survey (demographics and nine-item child food security measure) and 24-hour dietary recall data were collected in Spanish. Dietary data were collected in person on three occasions using a multiple-pass approach; nutrient intakes were calculated with NDS-R software. Separate multiple regression models were individually fitted for total energy, protein, dietary fiber, calcium, vitamin D, potassium, sodium, Vitamin C, and percentage of calories from fat and added sugars. Thirty-two children (64%) reported low or very low food security. Few children met the recommendations for calcium, dietary fiber, and sodium; and none for potassium or vitamin D. Weekend intake was lower than weekday for calcium, vitamin D, potassium, and vitamin C; and higher for percent of calories from fat. Three-day average dietary intakes of total calories, protein, and percent of calories from added sugars increased with declining food security status. Very low food security was associated with greater intakes of total energy, calcium, and percentage of calories from fat and added sugar. Chapter 10 not only emphasizes the alarming rates of food insecurity for this Hispanic subgroup, but describes the associations for food insecurity and diet among this sample of Mexican-origin children. Child-reported food insecurity situations could serve as a screen for nutrition problems in children. Further, the National School Lunch and School Breakfast Programs, which play a major beneficial role in children's weekday intakes, may not be enough to keep pace with the nutritional needs of low and very low food secure Mexican-origin children.

Interventions that cultivate sustainable food systems to promote health, prevent obesity, and improve food security have the potential for many large-scale and long-lasting benefits including improvements in social, environmental, health, and economic outcomes. The authors of chapter 11 briefly summarize findings from previous research examining associations between obesity and food insecurity and discuss the need for greater synergy between food insecurity initiatives and national obesity prevention public health goals in the United States. The common ground between these two nutrition-related public health issues is explored, and the transformation needed in research and advocacy communities around the shared goal of improving population health through individual, environmental, and policy level changes to promote healthy sustainable food systems is discussed. The authors of chapter 11 propose an ecological framework to simultaneously consider food insecurity and obesity that identifies levers for change to promote sustainable food systems to improve food security and prevent obesity.

The author of chapter 12 argues that understanding hunger and malnutrition requires an examination of what systems and institutions hold power over food. The concept of "food security" captures the notion of hunger not as a deficit of calories, but as a violation of a broader set of social, economic, and physical conditions. Gender is key to food insecurity and malnourishment, because women and girls are disproportionately disempowered through current processes and politics of food's production, consumption, and distribution. La Via Campesina has advocated for food sovereignty, through which communities have the right to define their own food and agriculture policy. Women's rights are central elements to food sovereignty. The author of chapter 12 suggests the role of the food industry demands attention within the food system, where power is concentrated in the hands of a few corporations.

Chapter 13 explores the power of large food corporations over the global food system. The authors of chapter 13 begin with the notion that global food systems are not meeting the world's dietary needs. About one billion people are hungry, while two billion people are overweight. India, for example, is experiencing rises in both: since 1995 an additional 65 million people are malnourished, and one in five adults is now overweight. This coexistence of food insecurity and obesity may seem like a paradox, but over- and undernutrition reflect two facets of malnutrition. The authors of chapter 13 believe that underlying both is a common factor: food systems are not driven to deliver optimal human diets but to maximize profits.

PART I
Defining Food Security and Insecurity

CHAPTER 1

Definitions of Food Security

United States Department of Agriculture

1.1 RANGES OF FOOD SECURITY AND FOOD INSECURITY

In 2006, USDA introduced new language to describe ranges of severity of food insecurity. USDA made these changes in response to recommendations by an expert panel convened at USDA's request by the Committee on National Statistics (CNSTAT) of the National Academies. Even though new labels were introduced, the methods used to assess households' food security remained unchanged, so statistics for 2005 and later years are directly comparable with those for earlier years for the corresponding categories.

1.1.1 USDA'S Labels Describe Ranges of Food Security

1.1.1.1 Food Security

- *High food security* (old label=Food security): no reported indications of food-access problems or limitations.
- *Marginal food security* (old label=Food security): one or two reported indications—typically of anxiety over food sufficiency or shortage of food in the house. Little or no indication of changes in diets or food intake.

USDA government document. Available online at http://www.ers.usda.gov/topics/food-nutrition-assistance/food-security-in-the-us/definitions-of-food-security.aspx.

1.1.1.2 Food Insecurity

- **Low food security** (old label=Food insecurity without hunger): reports of reduced quality, variety, or desirability of diet. Little or no indication of reduced food intake.
- **Very low food security** (old label=Food insecurity with hunger): Reports of multiple indications of disrupted eating patterns and reduced food intake.

1.2 CNSTAT REVIEW AND RECOMMENDATIONS

USDA requested the review by CNSTAT to ensure that the measurement methods USDA uses to assess households' access—or lack of access—to adequate food and the language used to describe those conditions are conceptually and operationally sound and that they convey useful and relevant information to policy officials and the public. The panel convened by CNSTAT to conduct this study included economists, sociologists, nutritionists, statisticians, and other researchers. One of the central issues the CNSTAT panel addressed was whether the concepts and definitions underlying the measurement methods—especially the concept and definition of hunger and the relationship between hunger and food insecurity—were appropriate for the policy context in which food security statistics are used.

1.2.1 The CNSTAT Panel

- Recommended that USDA continue to measure and monitor food insecurity regularly in a household survey.
- Affirmed the appropriateness of the general methodology currently used to measure food insecurity.
- Suggested several ways in which the methodology might be refined (contingent on confirmatory research). ERS has recently published Assessing Potential Technical Enhancements to the U.S. Household Food Security Measures and is continuing to conduct research on these issues.

The CNSTAT panel also recommended that USDA make a clear and explicit distinction between food insecurity and hunger.

- Food insecurity—the condition assessed in the food security survey and represented in USDA food security reports—is a household-level economic and social condition of limited or uncertain access to adequate food.
- Hunger is an individual-level physiological condition that may result from food insecurity.

The word "hunger," the panel stated in its final report, "...should refer to a potential consequence of food insecurity that, because of prolonged, involuntary lack of food, results in discomfort, illness, weakness, or pain that goes beyond the usual uneasy sensation." To measure hunger in this sense would require collection of more detailed and extensive information on physiological experiences of individual household members than could be accomplished effectively in the context of the CPS. The panel recommended, therefore, that new methods be developed to measure hunger and that a national assessment of hunger be conducted using an appropriate survey of individuals rather than a survey of households.

The CNSTAT panel also recommended that USDA consider alternative labels to convey the severity of food insecurity without using the word "hunger," since hunger is not adequately assessed in the food security survey. USDA concurred with this recommendation and, accordingly, introduced the new labels "low food security" and "very low food security" in 2006.

1.3 CHARACTERISTICS OF HOUSEHOLDS WITH VERY LOW FOOD SECURITY

Conditions reported by households with very low food security are compared with those reported by food-secure households and by households with low (but not very low) food security in the following graph:

The defining characteristic of very low food security is that, at times during the year, the food intake of household members is reduced and their normal eating patterns are disrupted because the household lacks money and other resources for food. Very low food security can be characterized in terms of the conditions that households in this category typically report in the annual food security survey.

- 98 percent reported having worried that their food would run out before they got money to buy more.

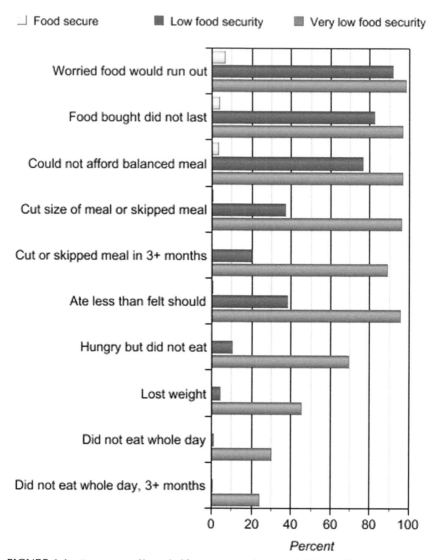

FIGURE 1.1 Percentage of households reporting indicators of adult food insecurity, by food security status, 2014.

Source: Calculated by ERS using data from the December 2014 Current Population Survey Food Security Supplement.

- 97 percent reported that the food they bought just did not last, and they did not have money to get more.
- 97 percent reported that they could not afford to eat balanced meals.
- 96 percent reported that an adult had cut the size of meals or skipped meals because there was not enough money for food.
- 89 percent reported that this had occurred in 3 or more months.
- 96 percent of respondents reported that they had eaten less than they felt they should because there was not enough money for food.
- 69 percent of respondents reported that they had been hungry but did not eat because they could not afford enough food.
- 45 percent of respondents reported having lost weight because they did not have enough money for food.
- 30 percent reported that an adult did not eat for a whole day because there was not enough money for food.
- 24 percent reported that this had occurred in 3 or more months.

All households without children that were classified as having very low food security reported at least six of these conditions, and 69 percent reported seven or more. Food-insecure conditions in households with children followed a similar pattern.

1.4   GLOBAL FOOD SECURITY

ERS provides quantitative and qualitative research and analysis on food security issues in developing countries, focusing on food security measurement and the key factors affecting food production and household access.

The annual ERS International Food Security Assessment is the only report to provide a 10-year projection of food security indicators in 76 low- and middle-income countries.

In recent years, the ERS International Food Security Assessments have estimated that between 500 and 700 million people in the 76 countries studied are food insecure. The estimate for 2015 is 475 million food-insecure people (food-insecure people are those consuming less than the nutritional target of about 2,100 calories per day). However, food security conditions vary from year to year because of changes in local food production and the financial capacity of countries to purchase food in global markets. Despite progress, Sub-Saharan Africa continues to account for the bulk of the food-insecure people in the countries assessed, followed by Asia and then Latin America and the Caribbean.

1.5 GLOSSARY

A **distribution gap** measures the difference between projected food consumption and the amount of food needed to increase consumption in food-deficit income groups within individual countries to meet nutritional requirements. Inadequate economic access to food is the major cause of chronic undernutrition in developing countries and is related to the level of income. In the ERS global food security model, the total projected amount of available food is allocated among different income groups using income distribution data.

A **nutrition gap** is estimated to measure food insecurity. This gap represents the difference between projected food supplies and the food needed to support per capita nutritional standards at the national level.

A **status quo gap** is estimated to measure changes in food security. This gap represents the difference between projected food supplies and the food needed to maintain per capita consumption of the most recent 3-year period.

PART II
Food Insecurity and Mental Health

Food Insecurity in Adults with Mood Disorders: Prevalence Estimates and Associations with Nutritional and Psychological Health

Karen M. Davidson and Bonnie J. Kaplan

2.1 BACKGROUND

Among health researchers, policy makers, practitioners, and decision makers, there are concerns about the growing global and ethical issue of food insecurity [1], defined as the limited or uncertain availability of nutritionally adequate and safe foods or limited or uncertain ability to acquire acceptable food in socially acceptable ways [2]. The impact of food insecurity on mental health is significant and may be attributed to its associations with suboptimal diet [3–5], and psychological issues such as depression, eating disorders, and impaired cognition [6–9]. Of the few investigations that have examined food insecurity in populations with confirmed diagnosis of a mental health condition, results have indicated an association with food insufficiency [10] and that patients in a psychiatric emergency unit who lacked food security had higher levels of psychological distress compared to food-secure individuals [11]. In a previous study of the determinants of food intakes in a sample of adults with confirmed mood disorders conducted by the authors [12], a high prevalence of inadequate intakes of several micronutrients (e.g., folate, vitamin B12, iron, zinc) was found. As part of this investigation, the authors also included food insecurity screening questions and measures of psychological symptoms, as it was speculated that food access would be associated with both dietary intake and mental function in this population. To further understanding of food insecurity in specific mental health populations, we used data from the study of adults with mood disorders to answer the following research questions: (1) what is the prevalence of food insecurity compared to a general population sample?; and (2) is food insecurity associated with poorer nutrient intakes and psychological function? It was hypothesized that in this sample of adults with mood disorders, food insecurity would be: (1) significantly more prevalent than in the general population; (2) associated with suboptimal nutrient intakes based on defined standards; and (3) associated with poorer psychological functioning.

2.2 METHODS

2.2.1 Sample

The data were from a cross-sectional nutrition survey of adults (>18 years) with clinically defined mood disorders (Structured Clinical Interview for DSM-IV-TR Axis I Disorders or SCID; [13]) who lived in the lower mainland of British Columbia; 146 individuals were randomly selected from the membership list of the Mood Disorder Association of British Columbia and invited to participate. Of the 146 randomly selected members, 26 were deemed ineligible and 23 refused to participate. Eleven of the 23 individuals who declined participation completed a non-response questionnaire, and statistical comparisons showed no significant differences compared to the respondents based on a variety of lifestyle habits (e.g., smoking, multivitamin use).

As part of the recruitment process, all individuals received phone follow-up from research staff to determine their interest and eligibility. Full details of recruitment, data collection, quality control, and overall nutrient intake results have been published previously [12]. This study adhered to guidelines in the Declaration of Helsinki and was approved by the University of Calgary's Conjoint Health Research Ethics Board. Sufficiency of statistical power (minimum 80%) to conduct the analysis was based on: (1) the hypothesis, food insecurity would be associated with lower nutrient intakes; (2) results from a Canadian study on food insecurity and nutrient intakes reporting mean intakes of folate of 424 and 378 micrograms for food-secure and food-insecure groups, respectively, with a common standard deviation of 18.5 [14]; and (3) an alpha level of 0.05 [15].

2.1.2 Measurements

2.1.2.1 Dependent Variable

2.1.2.1.1 Food Insecurity

Food insecurity status was determined as having an affirmative response to at least one of the two screening questions:

1. In the past 12 months did you worry that there would not be enough to eat because of lack of money? (Question description: Worry about food access).
2. In the past 12 months did you not have enough food to eat because of lack of money? (Question description: Compromised diet).

These two screening questions, which have previously been used in Statistics Canada national surveys [16], were also used in the British Columbia Nutrition Survey and allowed for direct comparisons of food insecurity between the sample and the general population. In the Canadian survey and in our study, a third question was also used: "In the past 12 months, did you or anyone in your household not have enough food to eat because of a lack of money?". All of our study respondents who responded affirmatively to question two also indicated "yes" to the third question. Due to this redundancy in responses, the third question was not included in the analysis.

2.1.2.2 Independent Variables

2.1.2.2.1 Psychological Functioning and Symptoms

Psychological, social, and occupational functioning was based on scores of the Global Assessment of Functioning Scale (GAF) [17]. Current symptoms of depression and mania were measured based on the Hamilton Depression Scale (Ham-D) [18], and Young Mania Rating Scale (YMRS) [19].

2.1.2.2.2   Diet

Nutrient intakes were derived from 3-day food records (two weekdays and one weekend day; non-consecutive) according to standard protocols [20]. The dietary variables included intakes of energy, macronutrients, and selected micronutrients (i.e., vitamins B6, B9, B12, and C, plus the trace minerals iron, and zinc) that function in cognition, and were analyzed using Food Processor SQL with the Canadian nutrient file [21].

2.1.3   Covariates

Covariates included in the analysis were body mass index (BMI; kg/m2; continuous variable) measured according to standard protocols used in national and provincial nutrition surveys [20], diagnosis of depressive or bipolar disorder based on the SCID (dichotomous), following a therapeutic (diet for a medical condition) diet (dichotomous), sociodemographic factors including age (continuous), gender (male/female), in a relationship or not (dichotomous), and income (dichotomized as low versus adequate, based on Canadian government standards).

2.1.3.1   Data Analysis

To test the first hypothesis, prevalence estimates of food insecurity (i.e., affirmative response to at least one of the two food insecurity screening questions) and low income were compared (using binomial tests of two proportions) to regional data available from the 1999 British Columbia Nutrition Survey (BCNS; n = 1,823) that contained the same income and food insecurity measures [20]. The BCNS was conducted by the British Columbia Ministry of Health Services, Health Canada, and the University of British Columbia. It randomly sampled British Columbians aged from 19 to 84 years. Most of the BCNS participants were married (65.2%), did not hold a university degree (85.5%), and had income levels above the poverty cut-off (74.8%).

To examine the second hypothesis, differences in nutrient intakes between respondents who were either food-insecure or secure were compared using Mann–Whitney U tests or binomial tests of two proportions based on the Dietary Reference Intakes [22]. Specifically, the proportion of the sample above and below the Estimated Average Requirements (EAR; to estimate prevalence of possible nutrient inadequacy) or within and above the Adequate Macronutrient

Distribution Ranges (AMDR; to estimate suboptimal or excess macronutrient intake) were analyzed. To test the third hypothesis, bivariate relationships between food insecurity and the psychological measures were analyzed using the Mann–Whitney U test and crude prevalence ratios. Cut-offs used to categorize the mental health measures included scores of 7 or less for the Ham-D [18] and 12 or less for the YMRS [19] which indicate that symptoms are currently absent. A cut-off of 60 for the GAF was used; scores less than this indicate more severe mental illness [23].

The final steps of the analysis involved conducting Poisson regression with robust variance to examine relationships between food insecurity and psychological functioning while adjusting for age, sex, relationship status, and income. This particular method was chosen as it is considered the most viable model option when analyzing cross-sectional data [24]. It provides adjusted prevalence ratios that are interpreted as the ratio of the proportion of the persons with the outcome (e.g., low GAF scores) over the proportion with the exposure (e.g., food insecurity). To test whether the Poisson model form fit the data, the goodness-of-fit Chi-squared test was applied. All data analyses were performed using Stata 10.0 [25].

2.3   RESULTS

2.3.1   Description of Sample

Similar to the general population sample studied in the BCNS, most respondents (n = 97) were females (n = 69; 71.1%). Unlike the BCNS, most did not have a university degree (n = 76; 78.4%) and were not in a relationship (61.9%). Approximately half of the sample (49.0%) had government defined low-income levels (Statistics Canada, 2013). Most (67.0%) had body mass indices greater than 24.9 kg/m2 and 19.0% indicated they were following a specific diet for medical reasons. Based on the SCID, 58 (59.8%) met criteria for bipolar disorder and 39 (40.2%) for depressive disorder. More than 85% of the participants were taking psychiatric medications.

2.3.2   Comparison of Prevalence Estimates of Food Insecurity

As hypothesized, the sample of adults with mood disorders had a higher proportion of individuals (36%) experiencing food insecurity (p < 0.001) compared to

TABLE 2.1　Indicators of low income and food insecurity compared to regional survey data.

Characteristic	Low-income status		Worry about food access[a]		Compromised diet[b]	
	Study	BCNS	Study	BCNS	Study	BCNS
	(n = 97)	(n = 1,823)	(n = 97)	(n = 1,823)	(n = 97)	(n = 1,823)
Prevalence estimates (%) with 95% confidence intervals for income and food insecurity						
Income and food insecurity	40.2 (30.4–50.7)*	25.2 (23.2–27.3)	3 6.1 (26.6 to 46.5)**	7.3 (6.1–8.6)	19.6 (12.2–28.9)**	3.4 (2.6–4.3)

* p < 0.05, ** p < 0.0001.

[a] Question: in the past 12 months did you worry that there would not be enough to eat because of lack of money?

[b] Question: in the past 12 months did you not have enough food to eat because of lack of money?

the general population sample (BCNS) (Table 2.1). The proportion of individuals reporting low-income status was also higher in adults with mood disorders ($p < 0.05$).

2.3.3   Food Insecurity and Nutrient Intakes

The respondents who reported food insecurity had lower median carbohydrate and vitamin C intakes (Table 2.2). The assessment of nutrient intakes based on the EARs indicated that those reporting food insecurity had higher proportions of inadequate protein, folate and zinc intakes ($p < 0.05$). The food-insecure group also had a higher proportion of participants with excess total fat intakes as measured by the AMDR ($p < 0.05$).

2.3.4   Food Insecurity and Psychological Functioning

Based on the bivariate analysis, some associations were found between food insecurity and psychological functioning when measured as continuous variables (Table 2.3). Ham-D and YMRS scores were significantly higher ($p < 0.05$) in participants with compromised diet (i.e., answered affirmatively to the food insecurity screening question "In the past 12 months did you not have enough food to eat because of lack of money"). In addition, higher depression scores were found in those reporting low income ($p < 0.05$).

Crude prevalence ratios (PRs) of food insecurity and the psychological measures (dichotomized based on their defined cut-off scores) showed significant association between food insecurity and mania symptoms (Table 2.4). When Poisson regression was applied, the ratio of the proportion of respondents with high YMRS scores over the proportion with food insecurity was 2.37 (95% CI 1.49–3.75, $p < 0.001$). Low income was also consistently associated with worse scores on all psychological measures (adjusted PRs ranged from 2.68 to 2.88, $p < 0.05$). The variables in the final three Poisson regression models included the psychological measures (i.e., GAF, YMRS, Ham-D, respectively) age, sex, and income. The goodness-of-fit Chi-squared test results were not significant for all models (goodness-of-fit 51.34 to 55.01, p's $> \chi 2\ (92) = 1.00$) suggesting that the Poisson model form fit the data.

TABLE 2.2 Comparisons of nutrient intakes by food security status

Measurement	Nutrient intakes according to food insecurity status[a] (median, IQR[b])		Nutrient intakes compared to Dietary Reference Intakes[c] and by food insecurity status[a]	
	Food insecure (n = 36)	Food secure (n = 61)	Food insecure (n = 36)	Food secure (n = 61)
Energy and macronutrients			% > AMDR[d] (95% CI)	
Energy (kilocalories)	2,406 (1,907; 2,875)	2,541 (1,802; 3,182)	—	—
Protein (% of calories)	14.0 (10.1; 18.4)	13.7 (9.4; 20.1)	5.5 (0.7–18.7)	0
Carbohydrates (% of calories)	47.8 (41.3; 57.3)*	52.9 (48.5; 59.8)	0.2 (0.1–19.5)*	11.4 (4.7–22.2)
Fibre (g)	21.5 (13.7; 29.4)	24.3 (15.3; 30.2)	—	—
Fats (% of calories)	35.4 (29.2; 45.0)*	34.2 (25.9; 38.9)	55.6 (38.1–72.1)*	29.5 (18.5–42.6)
			% < EARe (95% CI)	
Protein (g/kg)	—	—	19.4 (6.5–32.4)*	4.9 (0.5–10.3)
Vitamins				
Vitamin B$_6$ or pyridoxine (mg)	1.5 (1.0; 2.0)	1.5 (1.2; 2.2)	30.6 (16.3–48.1)	21.3 (11.9–33.7)
Vitamin B$_9$ or folate (dietary folate equivalents)[f]	74.2 (39.7; 204.4)	114.5 (55.0; 225.4)	83.3 (67.2–93.6)*	52.5 (39.3–65.4)
Vitamin B$_{12}$ or cobalamin (µg)	3.3 (1.9; 4.6)	3.5 (1.9; 5.1)	27.8 (14.2–45.2)	26.2 (15.8–39.1)
Vitamin C (mg)	80.5 (51.8; 180.7)*	128.8 (70.2; 216.7)	27.8 (14.2–45.2)	21.3 (11.9–33.7)

TABLE 2.2 (*Continued*)

Measurement	Nutrient intakes according to food insecurity status[a] (median, IQR[b])		Nutrient intakes compared to Dietary Reference Intakes[c] and by food insecurity status[a]	
	Food insecure (n = 36)	Food secure (n = 61)	Food insecure (n = 36)	Food secure (n = 61)
Trace minerals				
Iron (mg)	14.4 (10.2; 18.3)	16.0 (12.0; 23.8)	13.9 (4.7–29.5)	6.6 (1.8–15.9)
Zinc (mg)	8.9 (5.8; 11.8)	8.8 (6.4; 12.8)	52.8 (35.5–69.6)*	31.1 (19.9–42.3)

* $p < 0.05$.

[a] Food-insecure or secure based on respondents who answered yes to either of the following questions: (1) in the past 12 months did you worry that there would not be enough to eat because of lack of money; or (2) in the past 12 months did you not have enough food to eat because of lack of; food-secure based on respondents who answered no to both questions.

[b] Interquartile range.

[c] Reference values that are quantitative estimates of nutrient intakes to be used for planning and assessing diets for healthy people.

[d] Adequate macronutrient distribution ranges.

[e] Estimated average requirement.

[f] Dietary folate equivalents (DFE): values that adjust for the differences in absorption of food folate and synthetic folic acid; 1 mcg of DFE = 0.6 mcg of folic acid from fortified food or as a supplement taken with a meal = 1 mcg food folate; 0.5 mcg of folic acid from a supplement taken on an empty stomach.

TABLE 2.3 Indicators of low income and food insecurity by psychological measures in adults with mood disorders ($n = 97$)

Psychological measure	Low-income status		Median and interquartile range				
			Worry about food access[a]		Compromised diet[b]		
	Yes (n = 39)	No (n = 58)	Yes (n = 35)	No (n = 62)	Yes (n = 19)	No (n = 78)	
Global Assessment of Functioning	60 (51; 70)	65 (55; 75)	10 (3; 16)	9 (4; 14)	3 (1; 5)	3 (1; 4)	
Hamilton Depression Scale	60 (51; 65)*	65 (60; 75)	9 (4; 16)	9 (4; 14)	3 (2; 7)*	2 (1; 4)	
Young Mania Rating Scale	58 (43; 68)	65 (55; 75)	10 (3; 19)	9 (4; 14)	4 (2; 7)*	2 (1; 4)	

*p < 0.05.

[a] Question: in the past 12 months did you worry that there would not be enough to eat because of lack of money?

[b] Question: in the past 12 months did you not have enough food to eat because of lack of money?

TABLE 2.4 Crude and adjusted prevalence ratios of food insecurity[a], psychological, and income measures

Psychological measure	Crude prevalence ratio	Adjusted prevalence ratiob (95% confidence interval, p value)	
		Food insecurity	Low income
GAF (<60)	1.49 (0.97–2.31, $p = 0.072$)	1.42 (0.88–2.28, $p = 0.148$)	2.68 (1.53–4.71, $p = 0.001$)
YMRS (>12)	4.72 (1.36–16.24, $p = 0.006$)	2.37 (1.49–3.75, $p = 0.000$)	2.71 (1.59–4.63, $p = 0.000$)
Ham-D (>7)	1.02 (0.71–1.48, $p = 0.880$)	0.94 (0.59–1.49, $p = 0.783$)	2.88 (1.64–5.06, $p = 0.000$)

[a] Answered yes to question 1 or 2: Question 1: In the past 12 months did you worry that there would not be enough to eat because of lack of money?; and Question 2: In the past 12 months did you not have enough food to eat because of lack of money?

[b] Prevalence ratios from final models that included food insecurity, age, sex, and income; income is included in table as the adjusted prevalence ratios were significant.

2.4 DISCUSSION

The results of this study indicated that food insecurity was more prevalent in adults with mood disorders when compared to a general population. Sample respondents who reported food insecurity had significantly lower intakes of carbohydrates and vitamin C. A higher proportion of the adults with mood disorders had potentially inadequate intakes of protein, folate, and zinc. Finally, food insecurity was consistently associated with worsened mania symptoms in adults with mood disorders.

This is the first study to report the relationships between food insecurity and selected nutrients in adults with mood disorders. As shown in other studies of food insecurity in different populations [14, 26–28], there is higher risk of inadequate intakes of essential nutrients which can have profound effects on mental health. Research conducted in this sample of adults with mood disorders has shown positive correlations between nutrient intakes and psychological function [29]. There are multiple interconnected ways in which food insecurity could contribute to poorer nutritional and psychological health, including through nutrient deficiencies and/or the body's physiological response to the immense stress from not having the resources to feed oneself and/or one's family [30–32]. When people experience food insecurity, they may be forced to consume food in socially unacceptable ways [33] which may affect psychological function directly or indirectly by altering nutrient intakes and metabolism. For example, reduced blood folate levels (that may be a result of compromised dietary folate intake due to food insecurity) may impair one-carbon metabolism, which has been implicated in the pathogenesis of psychiatric symptoms [34]. While previous studies have shown links between food insecurity and psychological symptoms such as depression, anxiety, and suicide ideation [6, 7, 28, 35, 36], our results are the first to show association with mania symptoms. One mechanism underlying this relationship may be due to suboptimal blood levels of vitamin C caused by either lower intakes and/or increased breakdown due to stress. With lower blood levels of vitamin C, disequilibrium between ascorbic acid and vanadium metabolism may occur and result in mania symptoms [37].

The consistent association between low-income status and worse scores on the psychological measures is not surprising as other studies have shown links with depression and low income [38]. However, to our knowledge, this study is the first to report associations with YMRS and GAF scores. While food insecurity is an independent determinant of health [39], cross-sectional surveys have also shown it is strongly linked with another health determinant, inadequate income [16, 40–42]. Unfortunately, the cross-sectional nature of these investigations and our study cannot determine the temporality of the relationships among food insecurity, low income, and mental health. However, longitudinal studies have shown that associations between food insecurity and depression are bidirectional [30, 43].

For several reasons, the results of this study should be interpreted with caution. Limitations include the modest sample size and that the food insecurity screening questions may not discriminate varying degrees of food access. However, previous studies indicate that short form measurement tools are appropriate to determine the subjective experiences of food insufficiency [44]. Measurement error between food insecurity and psychological functioning may be present as the two may be correlated (e.g., people in depressed mood states might report increased food insecurity regardless of their true status) and thereby yield a spuriously stronger positive association. However, the multivariate analysis did not show an association between food insecurity and Ham-D scores. Finally, most of the sample was taking psychiatric medications which may impact on food-related behaviours (e.g., increase appetite) [45] and therefore, increase food insecurity.

2.5 CONCLUSIONS

Food insecurity is an important determinant of health and significant public health and social issue [39, 40]. This study extends the current body of food insecurity knowledge by providing prevalence estimates and reported associations with poorer nutrient intakes and psychological functioning in a sample of adults with mood disorders. While program and policy developments that strengthen food security would reduce health and social inequities [41, 42], it may also be speculated that such interventions may enhance the nutritional and psychological well-being of those with mood disorders. Future investigations should focus on interventions aimed at improving food security to establish its role in enhancing mental health.

REFERENCES

1. Food and Agriculture Organization (FAO) (2012) The state of food insecurity in the world: economic growth is necessary but not sufficient to accelerate reduction of hunger and malnutrition. FAO, Rome.
2. Anderson SA (1990) Core indicators of nutritional state for difficult-to-sample populations. J Nutr 120(Suppl 11):1559–1600
3. Monsivais P, Drewnowski A (2007) The rising cost of low-energy-density foods. J Am Diet Assoc 107(12):2071–2076
4. Klesges LM, Pahor M, Shorr RI, Wan JY, Williamson JD, Guralnik JM (2001) Financial difficulty in acquiring food among elderly disabled women: results from the women's health and aging study. Am J Public Health 91(1):68–75
5. Vozoris NT, Tarasuk VS (2003) Household food insufficiency is associated with poorer health. J Nutr 133(1):120–126
6. Alaimo K, Olson CM, Frongillo EA (2002) Family food insufficiency, but not low family income, is positively associated with dysthymia and suicide symptoms in adolescents. J Nutr 132(4):719–725
7. Wu Z, Schimmele C (2005) Food insufficiency and depression. Soc Perspect 48(4):481–504
8. Gunderson C, Kreider B (2009) Bounding the effects of food insecurity on children's health outcomes. J Health Econ 28(5):971–983
9. Sorsdahl K, Slopen N, Siefert K, Seedat S, Stein DJ, Williams DR (2011) Household food insufficiency and mental health in South Africa. J Epidemiol Community Health 65(5):426–431
10. Muldoon KA, Duff PK, Fielden S, Anema A (2013) Food insufficiency is associated with psychiatric morbidity in a nationally representative study of mental illness among food insecure Canadians. Soc Psychiatry Psychiatr Epidemiol 48(5):795–803
11. Gsisaru N, Kaufman R, Mirsky J, Witztum E (2011) Food insecurity and mental health: a pilot study of patients in a psychiatric emergency unit in Israel. Community Ment Health J 47(5):513–519

12. Davison KM, Kaplan BJ (2011) Vitamin and mineral intakes in adults with mood disorders: comparisons to nutrition standards and associations with sociodemographic and clinical variables. J Am Coll Nutr 30(6):547–558

13. First MB, Spitzer RL, Gibbon M, Williams JBW (2001) Structured clinical interview for DSM-IV-TR Axis I disorders, research version, non-patient edition. (SCID-I/NP). Biometrics Research, New York

14. Kirkpatrick SI, Tarasuk V (2008) Food insecurity is associated with nutrient inadequacies among Canadian adults and adolescents. J Nutr 138(3):604–612

15. Rosner B (2010) Fundamentals of biostatistics, 7th edn. Duxbury Press, Pacific Grove

16. Che J, Chen J (2001) Food insecurity in Canadian households [1998/99 data]. Health Rep 12(4):11–22

17. Endicott J, Spitzer RL, Fleiss JL, Cohen J (1976) The Global Assessment Scale: a procedure for measuring overall severity of psychiatric disturbance. Arch Gen Psychiatry 33(6):766–771

18. Hamilton M (1960) A rating scale for depression. J Neurol Neurosurg Psych 23:56–62

19. Young RC, Biggs JT, Ziegler VE, Meyer DA (1978) A rating scale for mania: reliability, validity and sensitivity. Br J Psychiatry 133:429–435

20. BC Ministry of Health Services (2004) British Columbia Nutrition Survey—Report on Energy and Nutrient Intakes. BC Ministry of Health Services, Victoria

21. Canada Health (2010) Canadian nutrient file. Health Canada, Ottawa

22. Institute of Medicine (2000) Dietary reference intakes: applications in dietary assessment. National Academies Press, Washington

23. Kessler RC, Barker PR, Colpe LJ, Epstein JF, Gfroerer JC, Hiripi E et al (2003) Screening for serious mental illness in the general population. Arch Gen Psychiatry 60(2):184–189

24. Lee J, Tan CS, Chia KS (2009) A practical guide for multivariate analysis of dichotomous outcomes. Ann Acad Med Singapore 38(8):714–719

25. StataCorp (2007) Stata Statistical Software: Release 10. StataCorp LP, College Station

26. McIntyre L, Glanville NT, Raine KD, Dayle JB, Anderson B, Battaglia N (2003) Do low-income lone mothers compromise their nutrition to feed their children? CMAJ 168(3): 686–691

27. Tarasuk V, McIntyre L, Li J (2007) Low-income women's dietary intakes are sensitive to the depletion of household resources in one month. J Nutr 137(8):1980–1987

28. Lee J, Frongillo EA Jr (2001) Nutritional and health consequences are associated with food insecurity among U.S. elderly persons. J Nutr 131(5):1503–1509

29. Davison KM, Kaplan BJ (2012) Nutrient intakes are correlated with overall psychiatric functioning in adults with mood disorders. Can J Psychiatry 57(2):85–92

30. Heflin CM, Siefert K, Williams DR (2005) Food insufficiency and women's mental health: findings from a 3-year panel of welfare recipients. Soc Sci Med 61(9):1971–1982

31. Huddleston-Casas C, Charigo R, Simmons LA (2009) Food insecurity and maternal depression in rural, low-income families: a longitudinal investigation. Public Health Nutr 12(8):1133–1140. doi:10.1017/S1368980008003650.226

32. Davison KM, Nig E, Chandrasekera U, Seely C, Cairns J, Mailhot-Hall L et al (2012) Promoting mental health through healthy eating and nutritional care. Dietitians of Canada, Toronto.

33. Piaseu N, Belza B, Shell-Duncan B (2004) Less money less food: voices from women in urban poor families in Thailand. Health Care Women Int 25(7):604–619

34. Kronenberg G, Colla M, Endres M (2009) Folic acid, neurodegenerative and neuropsychiatric disease. Curr Mol Med 9(3):315–323

35. Davison KM, Marshall-Fabien GL, Tecson A (2015) Association of moderate and severe food insecurity with suicidal ideation in adults: national survey data from three Canadian provinces. Soc Psychiatry Psychiatr Epidemiol 50(6):963–972. doi:10.1007/s00127-015-1018-1

36. Weiser SD, Young SL, Cohen CR, Kushel MB, Tsai AC, Tien PC et al (2011) Conceptual framework for understanding the bidirectional links between food insecurity and HIV/AIDS. Am J Clin Nutr 94(6):1729S–1739S

37. Naylor GJ, Corrigan FM, Smith AH, Connelly P, Ward NI (1987) Further studies of vanadium in depressive psychosis. Br J Psychiatry 150:656–661

38. Topuzoğlu A, Binbay T, Ulaş H, Elbi H, Aksu Tanık F, Zağlı N et al (2015) The epidemiology of major depressive disorder and subthreshold depression in Izmir, Turkey: prevalence, socioeconomic differences, impairment and help-seeking. J Affect Disord 181:78–86. doi:10.1016/j.jad.2015.04.017

39. McIntyre L, Rondeau K (2009) Food insecurity. In: Raphael D (ed) Social determinants of health: Canadian perspectives. Canadian Scholar's Press Inc, Toronto, Ontario

40. Tarasuk V, Mitchell A, Dachner N (2013) Household food insecurity in Canada, 2012. Research to identify policy options to reduce food insecurity (PROOF): 2014. Toronto, Ontario. Retrieved from http://www.nutritionalsciences.lamp.utoronto.ca

41. Office of Nutrition Policy & Promotion, Health Canada (2007) Canadian community health survey, cycle 2.2, nutrition (2004): income-related household food security in Canada. Health Canada, Ottawa

42. McIntyre L, Connor SK, Warren J (2000) Child hunger in Canada: results of the 1994 National Longitudinal Survey of Children and Youth. CMAJ 163(8):961–965

43. Huddleston-Casas C, Charigo R, Simmons LA (2008) Food insecurity and maternal depression in rural, low-income families: a longitudinal investigation. Public Health Nutr 12(8):1133–1140. doi:10.1017/S1368980008003650

44. Nord M, Andrews M, Carlson S (2008) Household food insecurity in the United States. USDA Economic Research Service, Washington, DC

45. Davison KM (2013) The relationships among psychiatric medications, eating behaviors, and weight. Eat Behav 14(2):187–191

Household Food Insecurity and Mental Distress Among Pregnant Women in Southwestern Ethiopia: A Cross Sectional Study Design

Mulusew G. Jebena, Mohammed Taha, Motohiro Nakajima, Andrine Lemieux, Fikre Lemessa, Richard Hoffman, Markos Tesfaye, Tefera Belachew, Netsanet Workineh, Esayas Kebede, Teklu Gemechu, Yinebeb Tariku, Hailemariam Segni, Patrick Kolsteren, and Mustafa al'Absi

3.1 BACKGROUND

Mental distress is becoming one of the major contributors to the global burden of diseases in developing countries [1, 2]. Mental distress is defined as a collection of mental problems that may not fall into standard diagnostic criteria and are characterized by symptoms of sleeplessness, exhaustion, irritability, poor memory, difficulty in concentrating, and somatic complaints [3].

Numerous epidemiologic studies have revealed that maternal distress during pregnancy is a public health concern in Sub Saharan African [4–8]. It has been

© Jebena et al. 2015. BMC Pregnancy and Childbirth, 2015, 15:250; DOI: 10.1186/s12884-015-0699-5. Distributed under the terms of the Creative Commons Attribution 4.0 International License (http://creativecommons. org/licenses/by/4.0/).

associated with low infant birth weight, impaired postnatal growth, increased frequency of infant diarrhea [9, 10], under nutrition, stunted growth, and poorer cognitive development [11, 12]. Previous study done in Ethiopia also showed that maternal distress was associated with child growth and low birth weight [10, 13].

If public health measures aim to improve the mental health status of pregnant women, there must be empirical data revealing the social, behavioral, economic and cultural factors associated with mental distress [14–17]. Among these, food insecurity is one of the determinants that are known to increase risk for mental distress [5, 18–21]. Food insecurity is defined as either the lack of nutritionally adequate and safe food or a limited ability to acquire acceptable food in socially acceptable ways [22]. It is a process that may start with household members being worried about not being able to access food followed by sacrificing the quality of their diet and then eventually reducing the amount of calories consumed. This usually occurs when regular access to adequate and nutritious food is limited [23, 24].

Food insecurity has been identified as a public health issue for women living in low-income households which results in poorer health outcomes [25]. For example, food insecurity is associated with decreased self-rated health status among adults and children [18, 26–28], depression and anxiety in mothers [29] reduced micronutrient intake, decreased fruit and vegetable consumption[30, 31], overweight [32], poor child physical growth [25, 33] low birth weight & elevated prenatal depressive symptoms [34]. The association between maternal depression and food insecurity has been also reported [5, 29–32]. But, few studies have examined the association between mental distress and food insecurity among pregnant women [35].

In Ethiopia both mental illness and food insecurity are reported. While previous research has shown an association between mental health and food insecurity among adults, non-pregnant women and children [7, 10, 13, 28, 36, 37], there is scant literature studying their relationship during pregnancy in Ethiopia. This study tried to answer the following two questions: One, what is the magnitude of both mental distress and food insecurity among pregnant women? Two is there an association between food insecurity and mental distress among pregnant women? Thus, the purpose of this study was to document the magnitude of food insecurity and mental distress and to examine their association during pregnancy. These findings can be used for program planners and researcher to further study the link between the two in southwestern Ethiopia.

3.2 METHODS

3.2.1 Study Setting

A facility-based cross-sectional study was conducted in Jimma Zone, one of the 20 administrative zones in Oromia Regional State, southwest Ethiopia. According to the Central Statistical Authority [38] 2.7 million people live in Jimma Zone on an area of 15,569 km² with a population density of 159.69 persons per km². Of these, 1.23 million are women. An estimated 31,050 women become pregnant every year and antenatal care coverage in the zone is 64.3 percent. There are three hospitals and 84 primary health centers in Jimma zone where pregnant mothers can receive antenatal care services. There are 12 public health facilities affiliated with Jimma university within the radius of 70 km for community based education program, research and services. Between the months of June and August 2013, a total of 2,987 pregnant women were on antenatal care (ANC) follow up at the 12 health facilities selected for this study [39].

3.2.2 Sample Size and Sampling

A single population proportion formula was used to estimate 660 pregnant women to be included in the sample. Assumptions for calculating the sample size were the degree of confidence interval (95%; $Z_{1-\alpha/2} = 1.96$), the estimated magnitude of mental distress among pregnant women ($P = 50\%$), a 4% degree of precision, and a non-response rate of 10%. A total of 11 health centers and one hospital were selected to sample the study subjects. These facilities were chosen purposefully based on their previous affiliation with the Jimma University Community Based Education Training Program (CBTP). Probability proportional to size sampling (PPS) techniques was employed to assign the number of pregnant women to be interviewed from each selected health center and hospital. All pregnant women coming for ANC services during the data collection period were taken as the source population for the study. Finally, a consecutive sampling technique was used to identify the study subjects from each of the health facilities until the desired sample size was obtained. Seriously ill pregnant women were excluded from the study and referred to the respective hospital.

3.2.3   Tools and Measurements

The instruments used for data collection were adapted from earlier studies and WHO guidelines [30, 40–48] and translated from English into the two most commonly spoken languages in the study setting (Afan Oromo and Amharic) by two fluent linguists (from the university Language and Literature Department). Each translation was then translated back by another person (linguists from the English Department) to ensure its consistency. Before the actual survey, the final translated questionnaires were pre tested on 5% of the sample at two different health institutions (one urban and the other rural) so that coherence, wording, sequencing and consistencies of all questions were amended accordingly. The result of this pretest was not included in the main analyses. Two days of intensive training on how to approaches the clients, interview techniques, ethical consideration and how to refer any mothers that needed help. Finally, exit interviews were conducted by trained data collectors at each of the 12 health facilities immediately after the mothers had received their ANC services. Field supervisors and the research team supervised the data collection process. Supervisors also checked the consistency of data before submission to the data manager. Ethical clearance was obtained from the Jimma University (JU) ethical review board. The participants were asked for their oral consent after the purpose of the study was clearly communicated. Confidentiality was ensured for each study participants.

3.2.4   Questionnaires to Measure Mental Distress

Mental distress was measured using The Self Reporting Questionnaire (SRQ-20). The SRQ-20 is a screening instrument developed by the World Health Organization (WHO) to assess the level of symptoms of overall mental distress one month preceding the survey [42]. Scores range from 0 to 20, with higher scores representing more severe mental distress. The SRQ has been used in several previous studies exploring the relationship between maternal psychological wellbeing and infant health [42, 46]. In developing countries, cutoff scores of ≥6, 7, 8, 4 and 10 have been used for identifying cases of mental distress [40, 42, 46, 48]. However, in this study we report the proportion of women scoring SRQ ≥7 or greater to indicate mental distress. The SRQ-20 has been validated in Ethiopia for measuring mental distress among rural pregnant women and the recommended cutoff points greater or equal to 7 have specificity of 62% and sensitivity of 68.4% [40]. If the scoring was <7, we coded "0" for no mental distress and if it was greater or equals to 7, we coded "1" to indicate presence of mental distress.

3.2.5   Questionnaire to Measure Household Food Insecurity Access Scale

Household food insecurity access was measured using items from the validated Household Food Insecurity Access Scale (HFIAS) that was specifically developed for use in developing countries [41, 43–45]. The HFIAS consists of 9 items specific to an experience of food insecurity occurring within the previous four weeks. Each respondent indicated whether they had encountered the following due to lack of food or money to buy food in the last one month: (1) worried about running out of food, (2) lack of preferred food, (3) the respondent or another adult had limited access to a variety of foods due to a lack of resources (4) forced to eat un preferred food due to lack of resources, (5) eating smaller portions, (6) skipping meals, (7) the household ran out of food, (8) going to sleep hungry, and (9) going 24 hours without food. Endorsed items are then clarified with reported estimates of the frequency of food insecurity (rarely, sometimes, and often). Scores range from 0 to 27 where higher scores reflect more severe food insecurity and lower scores represent less food insecurity. To determine the status of food insecurity the average HFIAS score (dividing the sum of Household score by number of household in the sample) was computed and then household food insecurity access prevalence (HFIAP) categories (food secure, mild, moderately and severely food insecure) was generated. But, since none of the mothers reported mild and severely food insecure households, HFIAP was only categorized in to two conditions [44].

3.2.6   Data Management and Analysis

The data were entered, cleaned, and analyzed using STATA version 12 for Microsoft Windows. Descriptive statistics, bivariate and multivariate logistic regression analyses were computed to examine the relationship between the explanatory variables and mental distress. Assuming a linear relationship between independent and dependent variables, the binary form of the dependent variable was coded as "1" for mental distress and "0" for not distressed. First binary logistic regression analyses were conducted between each and separate explanatory variables with the outcome of our interest (mental distress) (Model I) and reported using crude Odds Ratios. Finally, all significant variables($P < 0.05$) during the bivariate analyses were chosen for multivariate logistic regression modeling using forward selection method to explore the association of food insecurity with mental distress by controlling for other confounding variables such as age, occupation, monthly income, and ownership of agricultural land (Model II). Adjusted odds ratios (AOR) and their 95% confidence intervals (CI) were presented as indicators of strength of association. A p-value of 0.05 or less was used to determine the cut-off points for statistical significance.

3.3   RESULTS

3.3.1   Socio Demographic Characteristics

A total of 642 pregnant women agreed to participate in the study resulting in a response rate of 97.3%. The mean ($\pm$ SD) age of the mother was 25.5($\pm$ 4.9). The majority of the women (67.1%) were between the ages of 20 and 29 years old. A total of 91% of them were married. Just over half of them were from urban areas (58.7%) and most (69.6%) of them were Muslim by religion. About one quarter of the women were illiterate (26.8%). Half of the women had completed primary school or were able to read and write (26.6% and 23.5% respectively). The mean ($\pm$SD) household size was 3.9 ($\pm$1.6) (Table 1).

TABLE 3.1   Socio demographic characteristics of respondents in Jimma Zone, southwest Ethiopia, 2013.

Variables	Categories	Number	Percentage
Residence	Urban	377	58.7
	Rural	265	41.3
Household size	≤4	435	67.8
	>4	207	32.2
Age of respondents	15-24	277	43.2
	25-34	318	49.5
	≥35	47	7.3
Occupation	Farmer	172	26.8
	Gov. employee	107	16.7
	Merchant	83	12.9
	Daily laborer	66	10.3
	Others[a]	214	33.3

TABLE 3.1   *(Continued)*

Variables	Categories	Number	Percentage
Marital status	Married	584	91.0
	Others[b]	58	9.0
Ethnicity	Oromo	494	76.9
	Dawuro	32	5.0
	Amhara	75	11.7
	Others[c]	41	6.4
Religion	Muslim	447	69.6
	Orthodox	141	22.0
	Others[d]	54	8.4
Educational status	Illiterate	172	26.8
	Read and write	151	23.5
	Primary school	171	26.6
	Secondary school	60	9.3
	College & above	88	13.7
Number of children <5	0	234	36.4
	1	395	61.5
Food insecurity status	2	13	2.0
	Food secure	584	91.0
	Food insecure	58	9.0

[a]Students, housewife, maid workers, [b]single, widowed, divorced. [c]Wolaita, Kullo,Tigray, Yam, [d]Catholics, protestant

3.3.2   Socioeconomic and Household Wealth

Just over one quarter of the women reported a household monthly income less than 1,000 Ethiopian Birr (ETB). One hundred seventy-two (26.8%) of women were involved in farming, and 274 (42.7%) of women owned agricultural land. Three hundred forty-six (53.9%) of the respondents did not have electricity and 65 (10.1%) of the respondents did not have the use of a latrine. Most respondents (89.1%) used firewood or other agricultural products as fuel for cooking. Water was provided by access to pipe water or tap water in 61.1% of the households while 39% depend on spring, river or well water. While most had a radio (72%) and a cell phone (77.3%) in the home, other household items such as television, land line phones and refrigerators were uncommon (30.2%, 6.2% and 11.8% respectively) (Table 2).

TABLE 3.2　Socioeconomic and household wealth of respondents, southwest Ethiopia, 2013.

Household effects	Category	Number	Percentage
Electricity	Yes	296	46.1
	No	346	53.9
Television	Yes	194	30.2
	No	448	69.8
Radio	Yes	462	72.0
	No	180	28.0
Telephone- fixed	Yes	40	6.2
	No	602	93.8
Mobile telephone	Yes	496	77.3
	No	146	22.7
Refrigerator	Yes	76	11.8
	No	566	88.2
Own agricultural Land	Yes	274	42.7
	No	368	57.3
Latrine	Yes	577	89.9
	No	65	10.1
Type of fuel	Electricity	18	2.8
	Wood & Agri crops.	572	89.1
	Charcoal	48	7.5
Source of water supply	Others (e.g.kerosene)	4	0.8
	Pipe water	195	30.4
	Tap water	197	30.7
	spring	202	31.5
	Dig well	23	3.6
Family-monthly income[e]	Protected river water	25	3.9
	≤500	70	10.9
	500-999	106	16.5
Transportation	≥1000	466	72.6
Bicycle	Yes	40	6.2
	No	602	93.8

TABLE 3.2   *(Continued)*

Household effects	Category	Number	Percentage
Animal drawn cart	Yes	34	5.3
Bajaj[f]	No	608	94.7
	Yes	13	2.0
	No	629	98.0
Car	Yes	7	1.1
	No	635	98.9

[e]National currency in Birr, 3/27/2014 exchange rate 1 US Dollar = 19.37 Birr, 1 Euro = 27.627 Birr f Bajaj is a three wheel drive motor bike owned by household used for transportation.

3.3.3   Prevalence of Household Food Insecurity and Mental Distress

Internal consistency of the Household Food Insecurity Access Scale was strong (Cronbach's alpha = 0.96). The reliability coefficient (Cronbach's Alpha) for the SRQ-20 was 0.858, showing very good internal consistency of the items. Fifty-eight (9%) of the respondents were living in food insecure households and twenty-two percent of the pregnant women had mental distress (22.4%). The prevalence of mental distress was higher among pregnant women living in a food insecure environment (48.3%) when compared to those living in food secured households (19.9%).

3.3.4   Association between Food Insecurity and Mental Distress

Table 3 reports the association between food insecurity and mental distress. Pregnant women living in urban areas were more likely to have mental distress when compared to their rural counter parts (COR = 3.01, 95% CI =1.19, 7.60). Lower monthly income (≤500 ETB) was associated with mental distress (COR = 1.94, 95% CI = 1.12, 3.35). Pregnant women living in food insecure households were more likely to have mental distress (COR = 3.77, 95% CI = 2.17, 6.55). Multivariate logistic regression modeling analysis using the forward selection method was used to investigate the independent association of food insecurity with mental distress. The result showed that pregnant women living in a food insecure household were 4.15 times more likely to have mental distress (AOR = 4.15, 95% CI = 1.67, 10.32). In addition, this study reveals mental distress is also associated with a family history of mental illness (AOR = 5.85, 95% CI = 2.23, 15.34) and intimate partner violence (AOR = 3.20, 95% CI = 1.43, 7.13).

TABLE 3.3   Factors associated with mental distress among pregnant women in Jimma Zone, southwest Ethiopia, 2013

Variables	Categories	COR(95% CI)	AOR(95% CI)
Residence	Urban	1.74(1.17,2.58)	3.01(1.19,7.60)
	Rural	1	1
HH size	>4	0.80 (0.53,1.20)	2.26(1.25,4.19)
	≤4	1	1
Age	≥35	1.82(0.69,4.77)	3.44(0.91,12.96)
	30-34	1.44(0.62,3.35)	2.22(0.71,6.91)
	25-29	1.49(0.67,3.27)	1.67(0.58,4.81)
	20-24	1.26(0.57,2.77)	1.53(0.55,4.30)
	15-19	1	1
Occupation	Gove. employee	0.81(0.40,1.62)	0.33(0.09,1.26)
	Merchant	1.82(0.95,3.46)	1.04(0.31,3.50)
	Daily laborer	2.51(1.29,4.85)	1.03(0.26,3.94)
	Others	2.14(1.29,3.55)	1.65(0.62,4.41)
	Farmer	1	1
Monthly income	≤500	1.94(1.12,3.35)	0.76(0.32,1.82)
	501-999	1.35(0.83,2.21)	0.96(0.44,2.07)
	≥1000	1	1
Marital status	In Marriage	0.37(0.21,0.64)	0.05(0.01,0.33)
	Not in marriage	1	1
Land ownership	Yes	0.71(0.47,1.08)	1.23(0.68,2.22)
	No	1	1
Previous Hx of IPV	Yes	3.92(2.38,6.46)	3.20(1.43,7.13)
	No	1	1

TABLE 3.3   *(Continued)*

Variables	Categories	COR(95% CI)	AOR(95% CI)
Current Hx of IPV	Yes	1.95(1.10,3.48)	0.48(0.17,1.31)
	No	1	1
Food insecurity	Food insecure	3.77(2.17,6.55)	4.15(1.67,10.32)
	Food secure	1	1
FamHx of mental disorder	Yes	3.90(2.10,7.28)	5.85(2.23,15.34)
	No	1	1
Currently chewed khat	Yes	1.35(0.86,2.12)	2.25(1.05,4.83)
	No	1	1

COR- Crude Odds Ratios, AOR- Adjusted Odds Ratio, IPV-Intimate Partner Violence.

3.4 DISCUSSION

This is a multi-center, facility-based cross sectional study conducted to determine the magnitude of food insecurity and its correlates with mental distress during pregnancy among women following antenatal care. Previously, there was limited information related to common mental distress among pregnant women in this study setting. Moreover, findings from previous studies on the general adult community, adolescents and children [14, 18, 27, 29, 33]. To the best of our knowledge, there has been limited study of antenatal mental distress in Ethiopia; one study that measured the prevalence of postnatal maternal and paternal symptoms among adult women revealed rates of 37% for anxiety and 19.9% for depression respectively [7, 10, 11]. Therefore, the present study addresses the information gap regarding the prevalence of food insecurity and its correlate with mental distress among pregnant women and therefore expands the scientific literature in this area.

In this study the current prevalence of food insecurity was 9% and mental distress was 22.4%. The magnitude of household food insecurity in this study was very small when compared with a similar previous study [36, 37]. Dibaba et al., 2013 noted that 41% of women in that study reported food insecurity [7]. There are three possible reasons that might explain this observed difference. The first reason is the difference in the versions of the HFIAS used as a measure of food insecurity. We used 9 items and other studies have used a 6-item measurement scale. The second reason is the difference in study design. The other studies used a cohort and community-based cross sectional sampling as well as a different study population (i.e. the other studies used pregnant women in a community based design and a female adolescent population). This study focused on pregnant women in a facility-based design. The other difference might be due to the seasonal effect on household food insecurity status, i.e. there are times when most households run out of food and food insecurity tends to be high during those seasons.

The magnitude of mental distress was 22.4% in the present study, which is approximately consistent with a previous study conducted in Ethiopian communities in which about 22.7% of those subjects reported mental distress [15, 49, 50]. A similar study in southwest Ethiopia reported the magnitude of ante-natal depression as 19.9%, which is within the range of findings reported from Sub-Saharan Africa and other developing countries [15, 16, 32, 43]. However, the rate of mental distress in the present study is higher than the 11.7% of adults prevalence reported from Addis Ababa [14], the 12% from southern Ethiopia [49], and the 17% rate reported from Butajira, south-central Ethiopia [13] and it was also very high when compared to the reported 12% prevalence of prenatal depression in developed countries [51]. The difference in prevalence may be due to the tools used to measure the psychological morbidity or the cut-off points used. Some of these studies used higher/lower cut-off points to measure SRQ. The difference in prevalence might also be due to a difference in the study populations. Pregnant women are challenged by different physiologic and psychological stressors [7, 52–55], and the differential presence of hormones during pregnancies might contribute to the reported distress [56]. However, the high prevalence of maternal depression in poor countries may be related to women's exposure to several depression-related risk factors [52–54]. For instance, in settings of socio-economic adversity, there is substantial evidence for a relationship between intimate partner violence, lack of social support as well as poverty, and maternal mental distress during their pregnancies [57–60]. This study also reports that pregnant women who experienced intimate partner violence have significantly associated maternal distress. This is consistent with studies from other developing regions where intimate partner violence during pregnancy was reported to be significantly associated with poor mental health symptoms [54, 55]. The present study also revealed that urban pregnant women were more likely to report mental distress compared to their rural counterparts. The pressure associated with urban living where there is high economic vulnerability and poor mental health status might explain this. For example, a study conducted in rural Malawi utilizing the SRQ-20 showed that social circumstances and type of employment among women influenced psychological conditions and their expression [61]. The authors recommend that further studies be conducted in order to explore the relationship between urbanization and mental distress.

The present study showed a significant association between food insecurity and mental distress. The magnitude of food mental distress was higher (48.3%) among pregnant women living in a food insecure household compared with those from food secure households (19.9%). Other studies have also indicated that the effects of food insecurity extend beyond the nutritional effects [10, 11, 18, 62]. This is consistent with other studies documenting a close association between food insecurity and mental distress [5–7, 18, 20, 22, 30] that showed food insecurity to be associated with depressive symptoms. The mechanism by which food insecurity leads to mental distress is difficult to determine using a cross sectional design. Nonetheless, available evidence suggests that there are biological pathways and psycho social pathways where food insecurity leads to mental distress [52]. In general, despite differences in study subject characteristics and different cut-off points used to define mental distress and food insecurity, our study findings is in general agreement with previous reports of research completed in Ethiopia [7, 36, 53]. The present study has some methodological strength. There was a high response rate; the study was also facilities-based, which helped us include study participants with varied socio economic and cultural backgrounds. This study could also give an insight for policy makers and program planners to device appropriate strategies for prevention of food insecurity and mental health. However, there are also some study limitations. First, given that this is a cross-sectional study, it is not possible to establish causal relationships. The cross-sectional nature of our data did not enable us to show the direction of the association between household food insecurity and mental distress. Therefore, prospective longitudinal studies are needed in the study setting among pregnant women to understand better the bidirectional relationship. Secondly, mental distress is based on self-report and thus is subject to non-systematic errors in recall and systematic non-disclosure. Both types of errors may have led to some misclassification in our study. Thirdly, study findings may not be generalized to the broader pregnant population since our study was limited to only women presenting to the ANC at health care facilities. In a country such as Ethiopia, where a significant proportion of pregnant women do not have antenatal follow-up, the findings of the present study cannot be generalized to the broader population of pregnant women in the community. There are also other exogenous variables responsible for mental distress but these were not assessed by this study. Other studies may be required to understand different factors responsible for the cause of mental distress among pregnant women at the individual, community and societal level. In addition, lack of qualitative data to further explore their relationship is the major limitation to this study. More importantly, while this study reflects the public health importance of food security and mental distress, to quantify the prevalence of mental distress we used a cut off ≥7; there are different cut offs used by other researchers. We chose our particular cut-off because previous validation studies in Ethiopia using this cut point yielded an optimum SRQ cut-off score to determine the case for mental distress with 68.4% sensitivity and 62% specificity. In addition, studying at health facilities may have skewed the responses to affirmative responses for attempts to elicit help, rather than a manifestation of emotional distress [40].

3.5 CONCLUSION

This study found a 9% prevalence of food insecurity and a 22.4% prevalence of mental distress among pregnant women. The prevalence of mental distress is reported to be 48.3% and 19.9% among food insecure and food secure households respectively. Moreover, the results showed that there was an association between food insecurity and mental distress. However, the mechanism by which food insecurity is associated with mental health was not clearly known. Thus, further study is recommended to determine if food insecurity during pregnancy produces mental distress or weather mental distress is a contributing factor in the development of food insecurity. Future research should also investigate the impact of food insecurity and mental distress on pregnancy outcomes, fetal & child growth and development.

REFERENCES

1. WHO | The world health report 2001 - Mental Health: New Understanding, New Hope. Available from: http://www.who.int/whr/2001/en/whr01_en.pdf?ua=1
2. Lund C. Mental health in Africa: findings from the Mental Health and Poverty Project. International Review of Psychiatry. 2010;22(6):547–9.
3. Goldberg DP, Huxley P. Common mental disorders: a bio-social model. London. New York: Tavistock/Routledge; 1992.
4. Patel V, Kleinman A. Poverty and common mental disorders in developing countries. Bull World Health Organ. 2003;81(8):609–15.
5. Cole GT SM. The effect of food insecurity on mental health: Panel evidence from rural Zambia. Social science & medicine (1982). 2011;73(7):1071–9.

6. Weaver LJ, Hadley C. Moving beyond hunger and nutrition: a systematic review of the evidence linking food insecurity and mental health in developing countries. Ecology of Food and Nutrition. 2009;48(4):263–84.

7. Dibaba Y, Fantahun M, Hindin MJ. The association of unwanted pregnancy and social support with depressive symptoms in pregnancy: evidence from rural Southwestern Ethiopia. BMC Pregnancy and Childbirth. 2013;13(1):135.

8. Bennett HA, Einarson A, Taddio A, Koren G, Einarson TR. Prevalence of depression during pregnancy: systematic review. Obstet Gynecol. 2004;3(4):698–709.

9. Ross J, Hanlon C, Medhin G, Alem A, Tesfaye F, Worku B, et al. Perinatal mental distress and infant morbidity in Ethiopia: a cohort study. Arch Dis Child Fetal Neonatal Ed. 2011;96(1):F59–64.

10. Borders AEB, Grobman WA, Amsden LB, Holl JL. Chronic stress and low birth weight neonates in a low-income population of women. Obstetrics & Gynecology. 2007;9((2, Part 1):331–8.

11. Cook JT, Frank DA, Berkowitz C, Black MM, Casey PH, Cutts DB, et al. Food insecurity is associated with adverse health outcomes among human infants and toddlers. J Nutr. 2004;134(6):1432–8.

12. Rahman A, Harrington R, Bunn J. Can maternal depression increase infant risk of illness and growth impairment in developing countries? Child Care Health Dev. 2002;28:51–6.

13. Medhin G, Hanlon C, Dewey M, Alem A, Tesfaye F, Lakew Z, et al. The effect of maternal common mental disorders on infant undernutrition in Butajira, Ethiopia: The P-MaMiE study. BMC Psychiatry. 2010;10(1):32.

14. Kebede D, Alem A, Rashid E. The prevalence and socio-demographic correlates of mental distress in Addis Ababa, Ethiopia. Acta Psychiatrica Scandinavica. 1999;100:5–10.

15. Damena T, Mossie A, Tesfaye M. Khat Chewing and Mental Distress: A Community Based Study, in Jimma City, Southwestern Ethiopia. Ethiop J Health Sci. 2011;1(1):37–45.

16. Faisal-Cury A, Menezes P, Araya R, Zugaib M. Common mental disorders during pregnancy: prevalence and associated factors among low-income women in São Paulo, Brazil : depression and anxiety during pregnancy. Arch Womens Ment Health. 2009;12(5):335–43.

17. Golbasi Z, Kelleci M, Kisacik G, Cetin A. Prevalence and Correlates of Depression in Pregnancy among Turkish Women. Matern Child Health J. 2009;14(4):485–91.

18. Heflin CM, Siefert K, Williams DR. Food insufficiency and women's mental health: Findings from a 3-year panel of welfare recipients. Social Science & Medicine. 2005;61(9):1971–82.

19. Webb P, Coates J, Frongillo EA, Rogers BL, Swindale A, Bilinsky P. Measuring Household Food Insecurity: Why It's So Important and Yet So Difficult to Do. J Nutr. 2006;136(5):1404S–8S.

20. Maes KC, Hadley C, Tesfaye F, Shifferaw S. Food insecurity and mental health: Surprising trends among community health volunteers in Addis Ababa, Ethiopia during the 2008 food crisis. Soc Sci Med. 2010;70(9):1450–7.

21. Whitaker RC, Phillips SM, Orzol SM. Food insecurity and the risks of depression and anxiety in mothers and behavior problems in their preschool-aged children. Pediatrics. 2006;118(3):e859–68.

22. Siefert K. Food Insufficiency and the Physical and Mental Health of Low-Income Women. Women &amp. Health. 2001;32:159–77.

23. Bickel G, Nord M, Price C, Hamilton W. Cook, J:Guide to measuring household food security: Revised 2000. Food and Nutrition Service: Alexandria, VA; 2000.

24. Wunderlich GS, Norwood JL. Food Insecurity and Hunger in the United States: An Assessment of the Measure Washington. DC: The National Academies Press; 2006. Available from: http://www.nap.edu/catalog/11578.html.

25. Carter KN, Lanumata T, Kruse K, Gorton D. What are the determinants of food insecurity in New Zealand and does this differ for males and females? Australian and New Zealand Journal of Public Health. 2010;34(6):602–8.

26. Olson CM. Food Insecurity in Women: A Recipe for Unhealthy Trade-offs. Topics in Clinical Nutrition. 2005;20(4):321–8.

27. Tarasuk VS. Household food insecurity with hunger is associated with women's food intakes, health and household circumstances. J Nutr. 2001;131(10):2670–6.

28. Ashman SB, Dawson G. Maternal depression, infant psychobiological development, and risk for depression: Mechanisms of risk and implications for treatment. Washington, DC, US: American Psychological Association; 2002. p. 37–58.

29. Belachew T, Hadley C, Lindstrom D, Gebremariam A, Michael KW, Getachew Y, et al. Gender differences in food insecurity and morbidity among adolescents in Southwest Ethiopia. Pediatrics. 2011;127(2):e398–405.

30. Hadley C, Patil CL. Food insecurity in rural Tanzania is associated with maternal anxiety and depression. Am J Hum Biol. 2006;18(3):359–68.

31. Laraia BA, Siega-Riz AM, Gundersen C, Dole N. Psychosocial factors and socioeconomic indicators are associated with household food insecurity among pregnant women. J Nutr. 2006;136(1):177–82.

32. Sorsdahl K, Slopen N, Siefert K, Seedat S, Stein DJ, Williams DR. Household food insufficiency and mental health in South Africa. J Epidemiol Community Health. 2011;65(5):426–31.

33. Rao S, Yajnik CS, Kanade A, Fall CHD, Margetts BM, Jackson AA, et al. Intake of micronutrient-rich foods in rural indian mothers is associated with the size of their babies at birth: pune maternal nutrition study. J Nutr. 2001;131(4):1217–24.

34. Mikkelsen TB, Osler M, Orozova-Bekkevold I, Knudsen VK, Olsen SF. Association between fruit and vegetable consumption and birth weight: a prospective study among 43,585 Danish women. Scand J Public Health. 2006;34(6):616–22.

35. Hromi-Fiedler A, Bermúdez-Millán A, Segura-Pérez S, Pérezscamilla-E R. Household food insecurity is associated with depressive symptoms among low-income pregnant Latinas. Maternal & Child Nutrition. 2011;7(4):421–30.

36. Hadley C, Tegegn A, Tessema F, Cowan JA, Asefa M, Galea S. Food insecurity, stressful life events and symptoms of anxiety and depression in east Africa: evidence from the Gilgel Gibe growth and development study. J Epidemiol Community Health. 2008;62(11):980–6.

37. Alemseged F, Haileamlak A, Tegegn A, Tessema F, Woldemichael K, Asefa M, et al. Risk factors for chronic non-communicable diseases at Gilgel Gibe field research center, Southwest Ethiopia: population based study. Ethiop J Health Sci. 2012;22(Spec Iss):19–28.

38. CSA. Ethiopia demographic and health survey. Addis Ababa, Ethiopia and Calverton, Maryland, USA: Central Statistical Authority of Ethiopia and ORC Macro; 2006.

39. Jimma Zone Health Report , Health report; 2012/13, Jimma, Ethiopia (un published document)

40. Hanlon C, Medhin G, Alem A, Araya M, Abdulahi A, Hughes M, et al. Detecting perinatal common mental disorders in Ethiopia: validation of the self-reporting questionnaire and Edinburgh postnatal depression scale. J Affect Disord. 2008;108(3):251–62.

41. Daniel Maxwell RC. Measuring food insecurity: can an indicator based on localized coping behaviors be used to compare across contexts? Food Policy. 2008;33(6):533–40.

42. Beusenberg M, Orley JH, Health WHOD of M. A User's guide to the self-reporting questionnaire (SRQ. 1994); Available from: http://apps.who.int/iris/bitstream/10665/61113/1/WHO_MNH_PSF_94.8.pdf

43. Frongillo EA, Nanama S. Development and validation of an experience-based measure of household food insecurity within and across seasons in northern Burkina Faso. J Nutr. 2006;136(5):1409S–19S.

44. Coates J. Anne Swindale and Paula Bilinsky: Household Food Insecurity Access Scale (HFIAS) for Measurement of Household Food Access: Indicator Guide (v. 3). Washington, D.C.: Food and Nutrition Technical Assistance Project, Academy for Educational Development; 2007.

45. Knueppel D, Demment M, Kaiser L. Validation of the household food insecurity access scale in rural Tanzania. Public Health Nutrition. 2010;13(03):360.

46. Stewart RC, Kauye F, Umar E, Vokhiwa M, Bunn J, Fitzgerald M, et al. Validation of a Chichewa version of the Self-Reporting Questionnaire (SRQ) as a brief screening measure for maternal depressive disorder in Malawi, Africa. Journal of Affective Disorders. 2009;112(1–3):126–34.

47. Freeman M, Seris N, Matabula E, Price M. An evaluation of mental health services in the South Eastern Transvaal. Johannesburg: Centre for Health Policy, University of the Witwatersrand; 1991.

48. Deyessa N, Berhane Y, Alem A, Hogberg U, Kullgren G. Depression among women in rural Ethiopia as related to socioeconomic factors: A community-based study on women in reproductive age groups. Scand J Public Health. 2008;36(6):589–97.

49. Awas M, Kebede D, Alem A. Major mental disorders in Butajira, southern Ethiopia. Acta PsychiatricaScandinavica, Supplement. 1999;99(397):56–64.

50. Kebede D, Alem A. Major mental disorders in Addis Ababa, Ethiopia. I. Schizophrenia, schizoaffective and cognitive disorders. Acta Psychiatrica Scandinavica. 1999;100:11–7.

51. Husain N, Cruickshank K, Husain M, Khan S, Tomenson B, Rahman A. Social stress and depression during pregnancy and in the postnatal period in British Pakistani mothers: A cohort study. Journal of Affective Disorders. 2012;140(3):268–76.

52. Ashiabi GS, O'Neal KK. A framework for understanding the association between food insecurity and children's developmental outcomes. Child Development Perspectives. 2008;2(2):71–7.

53. Collins NL, Dunkel-Schetter C, Lobel M, Scrimshaw SC. Social support in pregnancy: Psychosocial correlates of birth outcomes and postpartum depression. Journal of Personality and Social Psychology. 1993;65(6):1243–58.

54. Curry MA, Perrin N, Wall E. Effects of abuse on maternal complications and birth weight in adult and adolescent women. Obstet Gynecol. 1998;92(4 Pt 1):530–4.

55. Mahenge B, Likindikoki S, Stöckl H, Mbwambo J. Intimate partner viol ence during pregnancy and associated mental health symptoms among pregnant women in Tanzania: a cross-sectional study. BJOG: Int J Obstet Gy. 2013;120(8):940–7.

56. Altemus M, Neeb CC, Davis A, Occhiogrosso M, Nguyen T, Bleiberg KL. Phenotypic differences between pregnancy-onset and postpartum-onset major depressive disorder. The Journal of Clinical Psychiatry. 2012;73(12):e1485–91.

57. Devries KM, Kishor S, Johnson H, Stöckl H, Bacchus L, Garcia-Moreno C, et al. Intimate partner violence during pregnancy: prevalence data from 19 countries. Reprod Health Matters. 2010;18:1–13.

58. Martin SL, Li Y, Casanueva C, Harris-Britt A, Kupper LL, Cloutier S. Intimate partner violence and women's depression before and during pregnancy. Violence Against Women. 2006;12:221–39.

59. Karmaliani R, Asad N, Bann CM, Moss N, McClure EM, Pasha O, et al. Prevalence of anxiety, depression and associated factors among pregnant women of Hyderabad, Pakistan. Int J Soc Psychiatry. 2009;55:414–24.

60. Ludermir AB, Lewis G, Valongueiro SA, de Araújo TVB, Araya R. Violence against women by their intimate partner during pregnancy and postnatal depression: a prospective cohort study. The Lancet. 2010;376:903–10.

61. Carta MG, Carpiniello B, Dazzan P, Reda MA. Depressive symptoms and occupational role among female groups: a research in a south-east African village. Psychopathology. 2000;33(5):240–5.

62. Harpham T, Huttly S, Silva MJD, Abramsky T. Maternal mental health and child nutritional status in four developing countries. J Epidemiol Community Health. 2005;59(12):1060–4.

PART III
Food Insecurity and HIV

Is Food Insecurity Associated with HIV Risk? Cross-Sectional Evidence from Sexually Active Women in Brazil

Alexander C. Tsai, Kristin J. Hung, and Sheri D. Weiser

4.1 INTRODUCTION

Since the early stages of the HIV epidemic, social science researchers have described how unequal gender relations and gendered structural constraints facilitate the spread of HIV among women [1], particularly among women in sub-Saharan Africa [2]–[5]. A series of newer studies have highlighted food insecurity as a central variable shaping women's risks of HIV exposure. Although women often occupy a primary role in household food production in sub-Saharan Africa, gender bias in the distribution of resources within the household places them at elevated risk for food insecurity compared with men [6],[7]. Qualitative research suggests that inadequate or uncertain access to food exerts an undue influence on women's decisions to engage in transactional sex or unprotected sex [8] or enter commercial sex work [9]. In a population-based study of women in Botswana and Swaziland, food insufficiency was associated with risky sexual behaviors including inconsistent condom use, even after statistical adjustment for education and household income [10],[11], and subsequent studies have replicated these findings in different settings in sub-Saharan Africa [12],[13].

Less is known about power relations, food insecurity, and sexual risk in Brazil, which has undergone successive changes in the gender and socio-geographic composition of its complex epidemic over the past three decades. Although the overall HIV incidence rate stabilized in the 1990s, this trend was driven primarily by reductions in new cases among men [14]. The number of new heterosexually transmitted infections among women has continued to increase, especially among women of reproductive age [14],[15]. Population-based data in Brazil suggest that knowledge about HIV prevention practices is well disseminated, but less than one-half of the population reports consistent condom use or condom use at last sexual intercourse (with a far greater proportion of men reporting condom use compared to women) [16],[17]. These differences are worrisome given that condom promotion has been given primary emphasis in Brazilian HIV prevention programming and policy [18] and that cities where the epidemic is most concentrated are generally characterized by the greatest inequalities between men and women [19]. Although some observers have hypothesized that unequal power relations between men and women in Brazil may explain the observed differences in condom use [16],[20],[21], little empirical work has been done to confirm this hypothesis [22].

At the country level, Brazil has achieved a very low score on the United Nations Development Programme's Gender Inequality Index relative to other countries that are considered to be advanced with regards to human development [23], and textbooks and didactic teaching tend to reinforce gender-based stereotypes [24]. Although increasingly gender-equitable legislation has been adopted in Brazil, such as mandatory joint titling of land to couples, landownership by men still exceeds that of women by a ratio of 8:1 [25]. In addition, violence against women is highly prevalent, particularly in the north and northeast regions of the country [26]. During times of economic adversity, women and girls living in resource-limited settings may experience worse nutritional and health outcomes than men and boys living in the same households [27]–[29]. These outcomes are relevant to the current context, given that Brazilian men exercise considerable decision-making dominance at the household level [30] and favor their sons in the distribution of resources within the household [31]–[33].

A significant methodological weakness of earlier studies linking food insecurity to HIV risk has been their reliance on non-validated measures of food insecurity and a lack of objective measures of nutritional risk [34], as well as failure to consider the specific mechanisms linking food insecurity to HIV risk reduction behaviors. To address these shortcomings, we analyzed data from a large, geographically diverse sample of women in Brazil to determine whether food insecurity is associated with condom use and/or symptoms of sexually transmitted infection, and to discern the mechanisms underlying these associations.

4.2 METHODS

4.2.1 Ethical Review

The data collection procedures for the 2006 Pesquisa Nacional de Demografia e Saúde da Criança e da Mulher (PNDS) were approved by the ICF Macro Institutional Review Board as well as by the Research Ethics Committee of the Sexually Transmitted Diseases/AIDS Reference and Training Center of the Health Secretariat of São Paulo state. All participants provided oral informed consent. Additional details on staff training, pretesting, and other survey procedures are detailed in the PNDS final report [35]. The specific analysis of PNDS data presented in this paper was reviewed by the Harvard School of Public Health Office of Human Research Administration and deemed exempt from full review because it was based on anonymous public-use data with no identifiable information on participants.

4.2.2 Data Source

The data for this study were drawn from the PNDS, a national study implemented by the Ministério da Saúde from March 11, 2006, to March 5, 2007, with technical assistance from ICF Macro and the US Agency for International Development. The PNDS employed a probabilistic, complex sampling design, and it was designed to be nationally representative of all women of reproductive age (i.e., 15–49 y). Of 17,411 eligible women selected, 15,575 were successfully interviewed, for a response rate of 89.5%. Data on the primary outcomes of interest were obtained only from women who were sexually active. Therefore, the analyses reported in this paper were restricted to women who reported sexual activity with a man in the previous 12 mo.

4.2.3   Conceptual Framework and Statistical Analysis

The conceptual framework guiding our analysis, adapted from previously published work [36], is depicted in Figure 1. Our primary focus was to explain HIV risk. Accordingly, the primary outcomes of interest in our analysis were (a) consistent condom use, defined as using a condom at each occasion of sexual intercourse in the previous 12 mo; (b) recent condom use, less stringently defined as using a condom with the most recent sexual partner; and (c) self-report of an itchy vaginal discharge in the previous 30 d, possibly indicating presence of a sexually transmitted infection. The primary explanatory variable of interest was household food insecurity, defined as "access by all people at all times to enough food for an active, healthy life" ([37], p. 1560). To measure food insecurity, we used the Escala Brasiliera de Segurança Alimentar (EBIA) [38],[39], a culturally adapted, Portuguese version of the US Household Food Security Survey Module [40]. Both the US and Brazilian scales differentiate between households with and without children in assessing the degree of food insecurity. The 18-item EBIA scale employs a recall period of 3 mo (compared to 12 mo for the US version) and has demonstrated good internal consistency as well as content, convergent, and internal validity [38],[39]. Using the previously validated algorithm [38],[39], we assigned participants into one of four categories: food secure, food insecure without hunger, moderately food insecure with hunger, and severely food insecure with hunger.

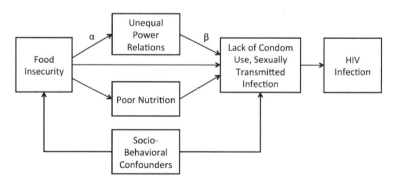

FIGURE 4.1   Conceptual framework linking food insecurity to HIV risk. The pathways between food insecurity and HIV risk may be direct or indirect. The indirect effects are mediated through unequal power relations and/or poor nutrition. For each hypothesized mediator, the indirect effect of the exposure on the outcomes (lack of condom use and sexually transmitted infection) is computed as the product of $\alpha \times \beta$. Other socio-behavioral variables may confound the observed association between food insecurity and HIV risk.

In our conceptual framework, food insecurity may have a direct effect on women's HIV risk, or the effect may be mediated through intervening variables that can also serve as programming or policy levers. In this analysis, we considered two specific mediators, poor nutrition and/or unequal power relations. Food insecurity may undermine women's ability to negotiate for condom use through its effects on nutritional risk and chronic energy deficiency. Previous research has also linked food insecurity to lack of control in sexual relations and to forced sex [10],[11]. In order to investigate these hypotheses, we operationalized chronic energy deficiency as a binary variable equal to one if the participant was underweight (defined as a body mass index <18.5 kg/m^2 [41]), zero otherwise. As a proxy for unequal power relations, we constructed a variable based on the participant's responses to questions about whether she felt it would be acceptable for a woman to refuse sexual intercourse with her husband or partner in five hypothetical scenarios [42]: if she knew he had a sexually transmitted disease, if she knew that he was having sexual intercourse with other women, if she had given birth to a child recently, if she was tired, or if she did not want to have sexual intercourse. Women who responded "no" to all five scenarios were (stringently) categorized as lacking control in sexual relations.

All analyses were conducted using the Stata statistical software package (version 12.0, StataCorp). To estimate the association between food insecurity and the outcomes of interest, we fit multivariable logistic regression models to the data, with cluster-correlated robust estimates of variance [43]–[45]. This modeling approach appropriately recognizes that variables measured at the level of the primary sampling unit have a smaller effective sample size and corrects the standard errors for potentially correlated observations between participants who live in the same primary sampling unit. We did not use the sampling weights provided by ICF Macro because this analysis was restricted to sexually active women only, and sampling weights were not provided for analyses restricted to this sub-sample of the population.

As depicted in the conceptual framework, socio-behavioral variables may confound the association between food insecurity and sexual risk. In our regression analyses, we therefore adjusted for potential confounding by the following socio-behavioral variables: age, racial/ethnic group (white [*branca*], black [*preta*], mixed [*parda*], Asian [*amarela*], or indigenous [*indigena*]), urban residence, macro-region of the country (north, northeast, southeast, south, or center-west), domestic partnership status (legally or formally married, not married but living with a partner in a consensual union, never married, separated, divorced, or widowed), Catholic religion, news reading frequency (reads the news daily, nearly every day, once per week, less than once per month, does not read), within-country quintiles of household asset wealth [46], and current use of cigarettes.

If simple linear regression models had been used in this analysis, we could have estimated the association between sexual risk and the hypothesized mediator adjusted for food insecurity (depicted as β in the conceptual framework), and then estimated the association between the food insecurity and the hypothesized mediator (α). The indirect effect of food insecurity on sexual risk, i.e., the portion of the effect of food insecurity on sexual risk that is due to the mediating variables of interest, could be computed as the product α×β [47], and the asymptotic variance would be computed using the multivariate delta method [48]. In the context of logistic regression, however, parameter rescaling tends to increase the apparent magnitude of the estimated regression coefficient and counters the effect of including the (potential) socio-behavioral confounders, as noted above. We therefore implemented a previously published algorithm [49],[50] to rescale the parameter estimates in order to decompose the total effect of food insecurity into its indirect and direct effects.

We also investigated whether the effect of food insecurity on the outcomes of interest varied according to domestic partnership status or fertility preferences. Fertility preferences were measured with a binary variable equal to one if the woman expressed a preference for no further childbearing and zero if the woman was undecided or expressed a preference to have more children. To assess potential effect modification, we included both the main effect terms and the interaction terms (with food insecurity) in the regression models and then used Wald-type F-tests to determine whether these variables modified the associations between food insecurity and the outcomes of interest. These interaction tests were based on our hypothesis that the adverse effects of food insecurity on condom use may be strongest among women who have a stronger preference for condom use, i.e., women who do not desire to bear more children or women who are not currently in a domestic partnership.

4.3   RESULTS

Of the 15,575 women interviewed for the study, 12,684 (81.4%) reported sexual activity with a man in the previous 12 mo and were therefore included in this analysis. The distributions of food insecurity scores were similar for women who were sexually active and women who were not: the mean EBIA scores were similar (t=0.18; p=0.86), and the percentages of women assigned to the different categories of food insecurity were also similar $(\chi^2=1.02; p=0.80)$. Consistent condom use was reported by 2,210 women (18.0%), condom use at last sexual intercourse was reported by 3,172 women (25.7%), and itchy vaginal discharge was reported by 1,337 women (10.8%). Summary statistics are presented in Table 1. On the EBIA, most women were categorized as food secure (9,343 [73.7%]), while 1,762 women (13.9%) were categorized as food insecure without hunger, 783 (6.2%) were categorized as moderately food insecure with hunger, and 473 (3.7%) were categorized as severely food insecure with hunger. The Cronbach's α for the EBIA was 0.91, indicating a high degree of internal consistency.

TABLE 4.1 Summary statistics

Variable	All Participants (n = 12,684)	Food Security Category		χ^2 Test Statistica
		Food Secure, Food Insecure without Hunger, or Moderately Food Insecure with Hunger (n = 12,211)	Severely Food Insecure with Hunger (n = 473)	
Age	32 (24–40)	32 (24–39)	34 (27–41)	20.6***
Race				90.4***
White (branca)	4,824 (39.4%)	4,725 (40.1%)	99 (21.2%)	
Black (preta)	1,210 (9.9%)	1,150 (9.8%)	60 (12.9%)	
Mixed (parda)	5,607 (45.8%)	55,334 (45.3%)	273 (58.6%)	
Asian (amarela)	342 (2.8%)	334 (2.8%)	8 (1.7%)	
Indigenous (indigena)	265 (2.2%)	239 (2.0%)	26 (5.6%)	
Urban residence	8,713 (70.5%)	8,432 (70.9%)	281 (59.4%)	29.0***
Macro-region				241.8***
North	2,126 (17.2%)	1,940 (16.3%)	186 (39.3%)	
Northeast	2,352 (19.0%)	2,218 (18.7%)	134 (28.3%)	
Southeast	2,627 (21.3%)	2,567 (21.6%)	160 (12.7%)	
South	2,713 (22.0%)	2,675 (22.5%)	38 (8.0%)	
Center-west	2,543 (20.6%)	2,488 (20.9%)	55 (11.6%)	
Domestic partnership status				68.1***
Other	2,779 (22.5%)	2,673 (22.5%)	106 (22.4%)	
Cohabiting	4,181 (33.8%)	3,944 (33.2%)	237 (50.1%)	

TABLE 4.1 *(Continued)*

Variable	All Participants (n = 12,684)	Food Security Category		χ² Test Statistica
		Food Secure, Food Insecure without Hunger, or Moderately Food Insecure with Hunger (n = 12,211)	Severely Food Insecure with Hunger (n = 473)	
Married	5,401 (43.7%)	5,271 (44.3%)	130 (27.5%)	
Catholic religion	10,248 (82.9%)	9,853 (82.9%)	395 (83.5%)	0.13
Frequency of reading the news				167.5***
Does not read	4,468 (36.2%)	4,174 (35.1%)	294 (62.2%)	
Less than once a month	2,363 (19.1%)	2,274 (19.1%)	89 (18.8%)	
At least once a week	2,773 (22.5%)	2,715 (22.9%)	58 (12.3%)	
Nearly every day	1,546 (12.5%)	1,527 (12.9%)	19 (4.0%)	
Daily	1,201 (9.7%)	1,188 (10.0%)	13 (2.8%)	
Household asset wealth index				550.5***
Most poor	2,501 (20.3%)	2,200 (18.7%)	281 (59.8%)	
Very poor	2,514 (20.4%)	2,398 (20.2%)	116 (24.7%)	
Poor	2,372 (19.2%)	2,319 (19.6%)	53 (11.3%)	
Less poor	2,517 (20.4%)	2,498 (21.1%)	19 (4.0%)	
Least poor	2,426 (19.7%)	2,425 (20.5%)	1 (0.2%)	
Smokes cigarettes	1,881 (15.2%)	1,768 (14.9%)	113 (23.9%)	28.7***

All data are number (percent), except for age, which is median (interquartile range).
ªRepresents the result of a non-parametric K-sample test on the equality of medians with continuity correction (for continuous variables) or Pearson's chi-squared test
(for categorical variables).
***Statistical significance at the level of $p<0.001$.
doi:10.1371/journal.pmed.1001203.t001

In multivariable analyses, severe food insecurity with hunger was associated with a statistically significant reduced odds of consistent condom use (adjusted odds ratio [AOR]=0.67; 95% CI, 0.48–0.92) and condom use at last sexual intercourse (AOR=0.84; 95% CI, 0.81–0.86) (Table 2). The estimated odds ratios for food insecurity categories of lesser severity did not have statistically significant associations with the condom use outcomes, whereas all categories of food insecurity were associated with increased odds of reporting symptoms

TABLE 4.2 Associations between food insecurity and sexual risk outcomes.

Variable	Consistent Condom Use in the Past 12 mo		Condom Use at Last Sexual Intercourse		Itchy Vaginal Discharge in the Past 30 d	
	OR (95% CI)	AOR (95% CI)	OR (95% CI)	AOR (95% CI)	OR (95% CI)	AOR (95% CI)
Food insecurity category						
Food secure	Ref	Ref	Ref	Ref	Ref	Ref
Food insecure without hunger	**0.85 (0.74–0.99)**	0.94 (0.79–1.10)	0.93 (0.82–1.04)	1.00 (0.87–1.14)	**1.61 (1.37–1.89)**	**1.46 (1.24–1.73)**
Moderately food insecure with hunger	0.88 (0.73–1.07)	1.00 (0.81–1.24)	0.86 (0.73–1.02)	0.95 (0.78–1.15)	**1.94 (1.58–2.37)**	**1.78 (1.44–2.22)**
Severely food insecure with hunger	**0.54 (0.40–0.72)**	**0.67 (0.48–0.92)**	**0.60 (0.47–0.77)**	**0.75 (0.57–0.98)**	**2.16 (1.67–2.79)**	**1.94 (1.47–2.56)**
Age (in 5-y blocks)	**0.77 (0.75–0.79)**	**0.93 (0.90–0.96)**	**0.75 (0.73–0.77)**	**0.84 (0.81–0.86)**	**0.95 (0.93–0.98)**	**0.95 (0.92–0.98)**
Race						
White	1.09 (0.98–1.22)	1.10 (0.97–1.24)	0.97 (0.88–1.07)	0.96 (0.86–1.08)	**0.81 (0.72–0.92)**	0.95 (0.83–1.10)
Black	1.09 (0.92–1.28)	1.06 (0.88–1.26)	1.03 (0.89–1.19)	0.97 (0.83–1.14)	0.96 (0.79–1.18)	0.98 (0.80–1.21)
Mixed	Ref	Ref	Ref	Ref	Ref	Ref
Asian	1.24 (0.95–1.63)	0.96 (0.71–1.30)	1.19 (0.93–1.53)	0.92 (0.68–1.24)	0.82 (0.56–1.21)	0.88 (0.58–1.33)
Indigenous	1.22 (0.87–1.70)	1.14 (0.83–1.57)	1.15 (0.84–1.57)	1.09 (0.80–1.49)	0.82 (0.53–1.27)	0.73 (0.45–1.18)
Urban residence	**1.75 (1.54–1.99)**	1.14 (1.00–1.30)	**1.89 (1.69–2.12)**	**1.31 (1.15–1.49)**	**0.78 (0.69–0.88)**	0.94 (0.82–1.08)
Macro-region						
North	Ref	Ref	Ref	Ref	Ref	Ref
Northeast	**0.78 (0.65–0.94)**	0.85 (0.71–1.02)	0.94 (0.80–1.12)	1.06 (0.89–1.25)	**0.68 (0.56–0.83)**	**0.69 (0.56–0.84)**
Southeast	0.87 (0.73–1.03)	**0.77 (0.64–0.92)**	0.86 (0.73–1.02)	0.85 (0.71–1.01)	**0.70 (0.58–0.85)**	0.87 (0.72–1.07)
South	0.84 (0.70–1.00)	**0.75 (0.62–0.91)**	0.87 (0.73–1.02)	0.87 (0.73–1.05)	**0.71 (0.59–0.86)**	0.91 (0.74–1.12)
Center-west	**0.81 (0.68–0.97)**	**0.76 (0.63–0.91)**	0.86 (0.73–1.02)	0.86 (0.72–1.02)	**0.77 (0.65–0.93)**	0.90 (0.75–1.08)

TABLE 4.2 (Continued)

Variable	Consistent Condom Use in the Past 12 mo		Condom Use at Last Sexual Intercourse		Itchy Vaginal Discharge in the Past 30 d	
	OR (95% CI)	AOR (95% CI)	OR (95% CI)	AOR (95% CI)	OR (95% CI)	AOR (95% CI)
Domestic partnership status						
Other	Ref	Ref	Ref	Ref	Ref	Ref
Cohabiting	**0.16 (0.15–0.19)**	**0.20 (0.18–0.23)**	**0.18 (0.16–0.20)**	**0.22 (0.19–0.25)**	**1.39 (1.20–1.63)**	**1.23 (1.05–1.44)**
Married	**0.15 (0.13–0.17)**	**0.19 (0.17–0.22)**	**0.14 (0.12–0.16)**	**0.19 (0.17–0.22)**	1.12 (0.96–1.31)	1.16 (0.98–1.37)
Catholic religion	**0.87 (0.77–0.97)**	1.00 (0.88–1.14)	**0.88 (0.79–0.98)**	1.04 (0.92–1.17)	0.93 (0.80–1.08)	0.94 (0.81–1.10)
Frequency of reading the news						
Does not read	Ref	Ref	Ref	Ref	Ref	Ref
Less than once a month	**1.60 (1.40–1.84)**	**1.34 (1.16–1.55)**	**1.54 (1.37–1.74)**	**1.29 (1.14–1.47)**	0.92 (0.79–1.07)	0.91 (0.77–1.06)
At least once a week	**1.91 (1.67–2.17)**	**1.38 (1.20–1.60)**	**1.86 (1.66–2.09)**	**1.39 (1.22–1.58)**	**0.71 (0.60–0.82)**	**0.78 (0.66–0.91)**
Nearly every day	**2.47 (2.13–2.86)**	**1.61 (1.36–1.91)**	**2.41 (2.11–2.75)**	**1.64 (1.41–1.91)**	**0.59 (0.49–0.73)**	**0.74 (0.60–0.91)**
Daily	**2.57 (2.19–3.01)**	**1.61 (1.33–1.94)**	**2.51 (2.17–2.90)**	**1.67 (1.41–1.98)**	**0.51 (0.40–0.64)**	**0.64 (0.50–0.84)**
Household asset wealth index						
Most poor	Ref	Ref	Ref	Ref	Ref	Ref
Very poor	**1.46 (1.23–1.72)**	**1.39 (1.15–1.68)**	**1.44 (1.26–1.66)**	**1.41 (1.20–1.66)**	0.88 (0.75–1.05)	1.01 (0.84–1.22)
Poor	**1.46 (1.26–1.71)**	**1.36 (1.13–1.65)**	**1.39 (1.21–1.60)**	**1.33 (1.12–1.57)**	**0.79 (0.66–0.94)**	1.02 (0.83–1.25)
Less poor	**1.61 (1.36–1.90)**	**1.46 (1.20–1.78)**	**1.46 (1.26–1.68)**	**1.39 (1.18–1.65)**	**0.80 (0.67–0.95)**	1.11 (0.90–1.37)
Least poor	**2.05 (1.74–2.42)**	**1.50 (1.21–1.86)**	**1.83 (1.59–2.11)**	**1.43 (1.18–1.73)**	**0.45 (0.37–0.55)**	**0.73 (0.56–0.94)**
Smokes cigarettes	0.96 (0.85–1.09)	0.93 (0.81–1.07)	**0.88 (0.79–0.99)**	**0.84 (0.74–0.96)**	0.90 (0.77–1.05)	**0.84 (0.71–0.98)**

Bold indicates statistical significance at the level of $p<0.05$.

OR, odds ratio; Ref, reference.

doi:10.1371/journal.pmed.1001203.t002

of itchy vaginal discharge. These estimated associations were also large in magnitude: evaluated at the mean of the other covariates, changing food security status from food secure to severely food insecure with hunger resulted in a change of the predicted probability of consistent condom use from 15.0% to 10.5%, while the predicted probability of self-reported itchy vaginal discharge changed from 9.2% to 16.4%. A number of other important patterns were also evident. There was little evidence of racial/ethnic differences. Condom use and self-reported itchy vaginal discharge were more likely among women in the north and northeast regions of the country, which have been less affected by the HIV epidemic and have experienced a slower rise in HIV incidence [14]. Consistent with previous work [51], greater reading frequency was associated with greater odds of condom use and reduced odds of self-reported itchy vaginal discharge.

In the mediation analyses, neither underweight nor lack of control in sexual relations proved to be substantive mediators of the relationship between food insecurity and the outcomes of interest (Table 3). Across the outcomes of interest, when these hypothesized mediators were included in the multivariable regression models, the estimated AOR for severe food insecurity changed minimally and even shifted away from the null. In addition, we assessed effect modification by domestic partnership status and fertility preferences. No statistically significant effect modification was observed for any of the outcomes, although severe food insecurity appeared to have the strongest association with condom use and symptoms of sexually transmitted infection among women who prefer more children.

4.4 DISCUSSION

Using data on 12,684 sexually active women sampled from diverse geographic regions of Brazil, we found that condom use was infrequent and that severe food insecurity with hunger was associated with reduced odds of condom use and increased odds of self-reported itchy vaginal discharge, possibly indicating presence of a sexually transmitted infection. These estimated associations were statistically significant, large in magnitude, and robust to statistical adjustment for known confounders. Given the infrequency of condom use among Brazilian women, a finding echoed by previous studies [16]–[18], and the centrality of condom promotion to Brazil's HIV prevention strategy, our findings have important implications for policy and programming for HIV prevention.

TABLE 4.3   Mediation and effect modification analyses.

Variable	Consistent Condom Use in the Past 12 mo	Condom Use at Last Sexual Intercourse	Itchy Vaginal Discharge in the Past 30d
Mediation analysis			
Unadjusted effect of severe food insecurity with hunger	**OR = 0.55** **(95% CI, 0.41–0.74)**	**OR = 0.62** **(95% CI, 0.48–0.79)**	**OR = 1.89** **(95% CI, 1.47–2.43)**
Adjusted effect[a]	**AOR = 0.68** **(95%CI, 0.49–0.94)**	**AOR = 0.76** **(95% CI, 0.57–0.99)**	**AOR = 1.52** **(95% CI, 1.16–2.01)**
Adjusted effect, accounting for underweight[a]	**AOR = 0.68** **(95% CI, 0.49–0.94)**	**AOR = 0.75** **(95% CI, 0.57–0.99)**	**AOR = 1.53** **(95% CI, 1.17–2.01)**
Percentage of total effect due to underweight	–0.9%	–2.7%	–0.8%
Adjusted effect, accounting for lack of control in sexual relations[a]	**AOR = 0.67** **(95% CI, 0.49–0.93)**	**AOR = 0.76** **(95% CI, 0.58–0.99)**	**AOR = 1.59** **(95% CI, 1.21–2.08)**
Percentage of total effect due to lack of control in sexual relations	–2.9%	–2.4%	–1.5%
Effect modification analysis			
By domestic partnership status			
Among non-partnered women	AOR = 0.68 (95% CI, 0.43–1.07)	AOR = 0.71 (95% CI, 0.47–1.08)	**AOR = 2.46** **(95% CI, 1.41–4.30)**
Among cohabiting women	AOR = 0.69 (95% CI, 0.39–1.21)	AOR = 0.75 (95% CI, 0.50–1.13)	**AOR = 1.60** **(95% CI, 1.07–2.39)**
Among married women	AOR = 0.44 (95% CI, 0.16–1.23)	AOR = 0.79 (95% CI, 0.41–1.53)	**AOR = 2.22** **(95% CI, 1.34–3.67)**
Wald-type F-test (p-value)	5.49 (0.48)	3.46 (0.75)	6.60 (0.36)

TABLE 4.3 *(Continued)*

Variable	Consistent Condom Use in the Past 12 mo	Condom Use at Last Sexual Intercourse	Itchy Vaginal Discharge in the Past 30 d
By fertility preference			
Among women who prefer no more children	AOR = 0.73 (95% CI, 0.46–1.16)	AOR = 0.76 (95% CI, 0.52–1.10)	AOR = 1.81 (95% CI, 1.14–2.88)
Among women who prefer more children/ undecided	**AOR = 0.54 (95% CI, 0.34–0.87)**	**AOR = 0.65 (95% CI, 0.43–0.98)**	**AOR = 2.03 (95% CI, 1.46–2.82)**
Wald-type F-test (*p*-value)	0.20 (0.90)	0.67 (0.72)	0.36 (0.83)

[a] Regression estimates adjusted for age, race, urban residence, macro-region, domestic partnership status, Catholic religion, news reading frequency, household asset wealth, and cigarette use. Bold indicates statistical significance at the level of *p* < 0.05.

doi:10.1371/journal.pmed.1001203.t003

It is well known that the social and economic marginalization of women constrains their ability to engage in HIV risk reduction behaviors [1]–[5]. Newer research has specifically identified food insecurity as a critical variable influencing women's risks of sexual violence [11] and exposure to HIV [10],[12],[13]. In these studies, however, food insecurity was measured using just one [10]–[12] or two [13] questions about food insufficiency. Food insecurity is a complex, multidimensional phenomenon characterized not only by insufficient food intake, but also by poor diet quality, disrupted eating patterns, and anxiety and uncertainty about access [52]. The single-question item incorporated into the US National Center for Health Statistics' Third National Health and Nutrition Examination Survey has demonstrated poor sensitivity for identifying food insecure households [53]. What is new about our analysis is our use of a well-developed, culturally adapted 18-item food insecurity scale that measures the entire range of human experience with food insecurity, from food security to severe food insecurity with hunger [38],[39].

We investigated two hypothesized mediators, underweight and lack of control in sexual relations, but these variables did not yield a substantive degree of mediation. This suggests that the observed association is due to a direct effect of food insecurity on sexual risk, or that it is due to mediation by unmeasured variables such as depression and other negative affect states [54] or condom use self-efficacy [55],[56]. In addition, measurement error in the variable for lack of control in sexual relations may have undermined our ability to adequately test for evidence of mediation. We did not have access to better developed measurements, such as the sexual relationship power scale [57]. However, even if the precise mechanism of action remains unknown, if the observed association is causal then food security interventions could still have beneficial effects on women's sexual risk, irrespective of the mechanism of action.

Our findings about food insecurity and sexual risk add to the burgeoning research base that highlights the importance of food insecurity as a variable of central importance in HIV prevention efforts [58],[59]. In Brazil, the specific targeting of high-risk women through food supplementation or livelihood interventions may help to equalize gender-based intra-household bargaining power differentials. Specifically, microfinance-based interventions have been promoted for reducing HIV risk [60],[61]. However, in Brazil, women have not been the traditional focus of microfinance initiatives as in other countries like India and Bangladesh [62].

Several limitations must be considered in interpreting our findings. First, we did not use the sampling weights provided by ICF Macro. Our analysis was restricted to sexually active women, a sub-sample for which sampling weights were not provided, and application of the weights for national representativeness could lead to unpredictable biases. This would not necessarily be considered as a source of potential bias, however, given that we do not attempt to generalize our findings to the population of Brazilian women of reproductive age. The large sample size does make our analysis to our knowledge the largest study of its kind to date, suggesting broad applicability across diverse socio-demographic groups in an emerging economy.

Second, measurement error in the outcomes of interest could bias our estimates in unpredictable ways. Although condom use at last sexual intercourse may be erroneously measured [63], consistent condom use signifies a greater degree of commitment and intention and is less subject to errors in reporting [64]. In addition, itchy vaginal discharge may be symptomatic of other conditions that are not sexually transmitted (e.g., genital/vulvovaginal candidiasis). In a meta-analysis of symptoms and signs of chlamydial infection and gonorrhea among women, the specificity of vaginal itching and vaginal discharge ranged from 65% to 79%, depending on the study setting [65]. Random measurement error in the dependent variable would have biased our estimates towards the null and resulted in more conservative estimates of association, however. In order for systematic measurement error in the dependent variable to bias our estimates away from the null, the systematic measurement error would need to be somehow related to the exposure of interest (food insecurity). Studies of food insecurity and HIV risk clearly warrant the collection of biomarker data, but these data are more difficult to obtain compared to measures of self-report. Because of the stigma attached to HIV, household surveys that incorporate HIV testing have typically experienced a 10%–20% lower response rate to HIV testing than to the household survey modules [66].

Third, study participants' mental health was not assessed. A cross-sectional study of women living in Goa, India, demonstrated that complaints of vaginal discharge were associated with both hunger and non-psychotic psychiatric morbidity [67],[68]. The study authors interpreted the latter finding as indicative of vaginal discharge as a bodily idiom of distress. Non-psychotic psychiatric morbidity could be potentially considered an unmeasured confounder with regards to our analysis. However, in light of prior studies linking food insecurity to depression and other markers of psychological distress [69]–[72], we believe that including such a variable in our regression models (if it had been available) would have resulted in over-adjustment by conditioning on part of the effect of interest.

Fourth, the observed associations between food insecurity and the outcomes could also be explained by unmeasured confounding. Our multivariable models included statistical adjustment for key variables known to confound the relationship between food insecurity and sexual risk (e.g., educational attainment and economic status) but may have omitted others. Most notably, we were unable to distinguish between consistent condom use with regular partners versus consistent condom use with casual partners [73]. There may have been differential patterning of food insecurity and condom use by partner type [74], given the highly negative meanings that may be attached to condom use in the context of marital relationships or regular sexual partnerships [1],[75]. Because casual or transactional partnerships are more likely to be economically motivated and characterized by greater frequency of HIV transmission risk behaviors [76],[77], failure to account for the type of partner could have confounded our estimates of the association between food insecurity and sexual risk. However, in our data we found no evidence of effect modification by domestic partnership status, and we would expect domestic partnership to be correlated (however weakly) with a lower propensity to have casual sexual partners.

Fifth, both the exposure and outcomes of interest were based on participant self-report. If study participants who provided responses consistent with more severe levels of food insecurity were also more likely to under-report condom use or over-report itchy vaginal discharge, this could have biased our estimates away from the null.

Sixth, the direction of causality is generically uncertain with data of a cross-sectional nature. However, the estimated associations presented in our analysis are strong, increasing with the intensity of the exposure, consistent with previously published research conducted in independent samples [10],[12],[13], plausible, and coherent with our socio-cultural understanding of the Brazilian context. Together these elements suggest our conservative interpretation of the data is correct [78], but longitudinal or experimental study designs in future work would help to strengthen claims of causality.

In summary, this study presents evidence from sexually active women living in Brazil that food insecurity is associated with reduced use of condoms during sexual intercourse. If the estimated association is causal, our findings suggest that interventions targeting food insecurity may have beneficial implications for HIV prevention. Individual-level cognitive and/or behavioral interventions targeting HIV risk avoidance or risk reduction behaviors are likely to be less than optimally effective if these structural factors are not also taken into account.

REFERENCES

1. Worth D (1989) Sexual decision-making and AIDS: why condom promotion among vulnerable women is likely to fail. Stud Fam Plann 20: 297–307.
2. Ulin PR (1992) African women and AIDS: negotiating behavioral change. Soc Sci Med 34: 63–73.
3. Schoepf BG (1988) Women, AIDS, and economic crisis in Central Africa. Can J Afr Stud 22: 625–644.
4. Ankrah EM (1991) AIDS and the social side of health. Soc Sci Med 32: 967–980.
5. Tsai AC, Subramanian SV (2012) Proximate context of gender-unequal norms and women's HIV risk in sub-Saharan Africa. AIDS 26: 381–386.
6. Quisumbing AR, editor. (2004) (2004) Household decisions, gender, and development: a synthesis of recent research. Washington (District of Columbia): International Food Policy Research Institute.
7. Tsai AC, Bangsberg DR, Emenyonu N, Senkungu JK, Martin JN, et al. (2011) The social context of food insecurity among persons living with HIV/AIDS in rural Uganda. Soc Sci Med 73: 1717–1724.
8. Miller CL, Bangsberg DR, Tuller DM, Senkungu J, Kawuma A, et al. (2011) Food insecurity and sexual risk in an HIV endemic community in Uganda. AIDS Behav 15: 1512–1519.
9. Oyefara JL (2007) Food insecurity, HIV/AIDS pandemic and sexual behaviour of female commercial sex workers in Lagos metropolis, Nigeria. SAHARA J 4: 626–635.
10. Weiser SD, Leiter K, Bangsberg DR, Butler LM, Percy-de Korte F, et al. (2007) Food insufficiency is associated with high-risk sexual behavior among women in Botswana and Swaziland. PLoS Med 4: e260. doi:10.1371/journal.pmed.0040260.
11. Tsai AC, Leiter K, Wolfe WR, Heisler M, Shannon K, et al. (2011) Prevalence and correlates of forced sex perpetration and victimization in Botswana and Swaziland. Am J Pub Health 101: 1068–1074.
12. Davidoff-Gore A, Luke N, Wawire S (2011) Dimensions of poverty and inconsistent condom use among youth in urban Kenya. AIDS Care 23: 1282–1290.
13. Cluver LD, Orkin M, Boyes M, Gardner F, Meinck F (2011) Transactional sex amongst AIDS-orphaned and AIDS-affected adolescents predicted by abuse and extreme poverty. J Acquir Immune Defic Syndr 58: 336–343.
14. Fonseca MG, Bastos FI (2007) Vinte e cinco anos da epidemia de AIDS no Brasil: principais achados epidemiológicos, 1980–2005. Cad Saude Publica 23: Suppl 3S333–S344.
15. Barbosa Junior A, Szwarcwald CL, Pascom AR, Souza Junior PB (2009) Tendências da epidemia de AIDS entre subgrupos sob maior risco no Brasil, 1980–2004. Cad Saude Publica 25: 727–737.
16. Szwarcwald CL, Barbosa-Junior A, Pascom AR, de Souza-Junior PR (2005) Knowledge, practices and behaviours related to HIV transmission among the Brazilian population in the 15–54 years age group, 2004. AIDS 19: Suppl 4S51–S58.
17. Brasil Ministerio da Saude, Secretaria de Vigilancia Saude (2005) Programa nacional de DST e Aids. Pesquisa de conhecimento attitudes e praticas na populacao Brasileira de 15 a 54 anos, 2004. Brasilia: Secretaria de Vigilancia Saude, Programa Nacional de DST e Aids, Ministerio da Saude.

18. Miranda AE, Figueiredo NC, McFarland W, Schmidt R, Page K (2011) Predicting condom use in young women: demographics, behaviours and knowledge from a population-based sample in Brazil. Int J STD AIDS 22: 590–595.

19. Grangeiro A, Escuder MM, Castilho EA (2010) A epidemia de AIDS no Brasil e as desigual-dades regionais e de oferta de servico. Cad Saude Publica 26: 2355–2367.

20. Hebling EM, Guimaraes IR (2004) Mulheres e AIDS: relações de gênero e uso do condom com parceiro estável. Cad Saude Publica 20: 1211–1218.

21. Villela WV, Doreto DT (2006) Sobre a experiencia sexual dos jovens. Cad Saude Publica 22: 2467–2472.

22. Chacham AS, Maia MB, Greco M, Silva AP, Greco DB (2007) Autonomy and susceptibility to HIV/AIDS among young women living in a slum in Belo Horizonte, Brazil. AIDS Care 19: Suppl 1 S12–S22.

23. United Nations Development Programme (2011) Human development report 2011: sus-tainability and equity—a better future for all. New York: United Nations Development Programme.

24. Pena MV, Correia M (2002) Brazil gender review: issues and recommendations. Report No. 23442-BR. Washington (District of Columbia): The World Bank.

25. Deere CD, Leon M (2003) The gender asset gap: land in Latin America. World Dev 31: 925–947.

26. Reichenheim ME, Moraes CL, Szklo A, Hasselmann MH, de Souza ER, et al. (2006) The magnitude of intimate partner violence in Brazil: portraits from 15 capital cities and the Federal District. Cad Saude Publica 22: 425–437.

27. Dercon S, Krishnan P (2000) In sickness and in health: risk sharing within households in rural Ethiopia. J Polit Econ 108: 688–727.

28. Behrman J (1988) Intrahousehold allocation of nutrients in rural India: Are boys favored? Do parents exhibit inequality aversion? Oxford Econ Papers 40: 32–54.

29. Rose E (1999) Consumption smoothing and excess female mortality in rural India. Rev Econ Stat 81: 41–49.

30. Ponczek V (2011) Income and bargaining effects on education and health in Brazil. J Dev Econ 94: 242–253.

31. Thomas D (1994) Like father, like son; like mother, like daughter: parental resources and child height. J Hum Resources 29: 950–989.

32. Emerson PM, Souza AP (2007) Child labor, school attendance, and intrahousehold gender bias in Brazil. World Bank Econ Rev 21: 301–316.

33. Rangel MA (2006) Alimony rights and intrahousehold allocation of resources: evidence from Brazil. Econ J 116: 627–658.

34. Rollins N (2007) Food insecurity—a risk factor for HIV infection. PLoS Med 4: e301. doi:10.1371/journal.pmed.0040301.

35. Ministerio da Saude (2008) PNDS 2006: Pesquisa Nacional de Demografia e Saude da Crianca e da Mulher. Brasilia: Ministerio da Saude.

36. Jewkes R (2010) HIV/AIDS. Gender inequities must be addressed in HIV prevention. Science 329: 145–147.

37. Anderson SA (1990) Core indicators of nutritional state for difficult-to-sample populations. J Nutr 120: 1559–1600.

38. Pérez-Escamilla R, Segall-Correa AM, Kurdian Maranha L, Sampaio MdFA, Marin-Leon L, et al. (2004) An adapted version of the U.S. Department of Agriculture Food Insecurity

module is a valid tool for assessing household food insecurity in Campinas, Brazil. J Nutr 134: 1923–1928.

39. Melgar-Quinonez HR, Nord M, Pérez-Escamilla R, Segall-Correa AM (2008) Psychometric properties of a modified US-household food security survey module in Campinas, Brazil. Eur J Clin Nutr 62: 665–673.

40. Carlson SJ, Andrews MS, Bickel GW (1999) Measuring food insecurity and hunger in the United States: development of a national benchmark measure and prevalence estimates. J Nutr 129: 510S–516S.

41. World Health Organization (2000) Obesity: preventing and managing the global epidemic. WHO Obesity Technical Report Series No. 894. Geneva: World Health Organization.

42. Pallitto CC, O'Campo P (2005) Community level effects of gender inequality on intimate partner violence and unintended pregnancy in Colombia: testing the feminist perspective. Soc Sci Med 60: 2205–2216.

43. Froot KA (1989) Consistent covariance matrix estimation with cross-sectional dependence and heteroskedasticity in financial data. J Financial Quant Anal 24: 333–355.

44. Williams RL (2000) A note on robust variance estimation for cluster-correlated data. Biometrics 56: 645–646.

45. Rogers WH (1993) Regression standard errors in clustered samples. Stata Tech Bull 13: 19–23.

46. Filmer D, Pritchett LH (2001) Estimating wealth effects without expenditure data—or tears: an application to educational enrollments in states of India. Demography 38: 115–132.

47. Baron RM, Kenny DA (1986) The moderator-mediator variable distinction in social psychological research: conceptual, strategic, and statistical considerations. J Pers Soc Psychol 51: 1173–1182.

48. Sobel ME (1982) Asymptotic confidence intervals for indirect effects in structural equation models. In: Leinhardt S, editor. Sociological methodology. Washington (District of Columbia): American Sociological Association.

49. Karlson KB, Anders H (2011) Decomposing primary and secondary effects: a new decomposition method. Res Social Strat Mobility 29: 221–237.

50. Breen R, Karlson KB, Holm A (2011) Total, direct, and indirect effects in logit models. Centre for Strategic Educational Research Working Paper No. 0005. Denmark: Danish School of Education, Aarhus University.

51. Jukes M, Simmons S, Bundy D (2008) Education and vulnerability: the role of schools in protecting young women and girls from HIV in southern Africa. AIDS 22: Suppl 4S41–S56.

52. Radimer KL, Olson CM, Campbell CC (1990) Development of indicators to assess hunger. J Nutr 120: Suppl 111544–1548.

53. Frongillo EA Jr, Rauschenbach BS, Olson CM, Kendall A, Colmenares AG (1997) Questionnaire-based measures are valid for the identification of rural households with hunger and food insecurity. J Nutr 127: 699–705.

54. Sikkema KJ, Watt MH, Drabkin AS, Meade CS, Hansen NB, et al. (2010) Mental health treatment to reduce HIV transmission risk behavior: a positive prevention model. AIDS Behav 14: 252–262.

55. Bandura A (1994) Social cognitive theory and the exercise of control over HIV infection. In: DiClemente R, Peterson J, editors. Preventing AIDS: theories and methods of behavioral interventions. New York: Plenum Press. pp. 25–59.

56. Wulfert E, Safren SA, Brown I, Wan CK (1999) Cognitive, behavioral, and personality correlates of HIV-positive persons' unsafe sexual behavior. J Appl Soc Psychol 29: 223–244.

57. Pulerwitz J, Gortmaker SL, DeJong W (2000) Measuring sexual relationship power in HIV/STD research. Sex Roles 42: 637–660.

58. Anema A, Vogenthaler N, Frongillo EA, Kadiyala S, Weiser SD (2009) Food insecurity and HIV/AIDS: current knowledge, gaps, and research priorities. Curr HIV/AIDS Rep 6: 224–231.

59. Weiser SD, Young SL, Cohen CR, Kushel MB, Tsai AC, et al. (2011) Conceptual framework for understanding the bidirectional links between food insecurity and HIV/AIDS. Am J Clin Nutr 94: 1729S–1739S.

60. Dworkin SL, Blankenship K (2009) Microfinance and HIV/AIDS prevention: assessing its promise and limitations. AIDS Behav 13: 462–469.

61. Pronyk PM, Hargreaves JR, Kim JC, Morison LA, Phetla G, et al. (2006) Effect of a structural intervention for the prevention of intimate-partner violence and HIV in rural South Africa: a cluster randomised trial. Lancet 368: 1973–1983.

62. Brusky B, Fortuna JP (2002) Understanding the demand for microfinance in Brazil: a qualitative study of two cities. Rio de Janeiro: National Development Bank of Brazil (BNDES).

63. Gallo MF, Behets FM, Steiner MJ, Hobbs MM, Hoke TH, et al. (2006) Prostate-specific antigen to ascertain reliability of self-reported coital exposure to semen. Sex Transm Dis 33: 476–479.

64. Hearst N, Chen S (2004) Condom promotion for AIDS prevention in the developing world: is it working? Stud Fam Plann 35: 39–47.

65. Sloan NL, Winikoff B, Haberland N, Coggins C, Elias C (2000) Screening and syndromic approaches to identify gonorrhea and chlamydial infection among women. Stud Fam Plann 31: 55–68.

66. Mishra V, Vaessen M, Boerma JT, Arnold F, Way A, et al. (2006) HIV testing in national population-based surveys: experience from the Demographic and Health Surveys. Bull World Health Organ 84: 537–545.

67. Patel V, Pednekar S, Weiss H, Rodrigues M, Barros P, et al. (2005) Why do women complain of vaginal discharge? A population survey of infectious and pyschosocial risk factors in a South Asian community. Int J Epidemiol 34: 853–862.

68. Patel V, Weiss HA, Kirkwood BR, Pednekar S, Nevrekar P, et al. (2006) Common genital complaints in women: the contribution of psychosocial and infectious factors in a population-based cohort study in Goa, India. Int J Epidemiol 35: 1478–1485.

69. Tsai AC, Bangsberg DR, Frongillo EA, Hunt PW, Muzoora C, et al. (2012) Food insecurity, depression and the modifying role of social support among people living with HIV/AIDS in rural Uganda. Soc Sci Med. doi:10.1016/j.socscimed.2012.02.033.

70. Weaver LJ, Hadley C (2009) Moving beyond hunger and nutrition: a systematic review of the evidence linking food insecurity and mental health in developing countries. Ecol Food Nutr 48: 263–284.

71. Heflin CM, Siefert K, Williams DR (2005) Food insufficiency and women's mental health: findings from a 3-year panel of welfare recipients. Soc Sci Med 61: 1971–1982.

72. Kim K, Frongillo EA (2007) Participation in food assistance programs modifies the relation of food insecurity with weight and depression in elders. J Nutr 137: 1005–1010.

73. Misovich SJ, Fisher JD, Fisher WA (1997) Close relationships and elevated HIV risk behavior: evidence and possible underlying psychological processes. Rev Gen Psychol 1: 72–107.

74. de Walque D, Kline R (2011) Variations in condom use by type of partner in 13 sub-Saharan African countries. Stud Fam Plann 42: 1–10.

75. Tavory I, Swidler A (2009) Condom semiotics: meaning and condom use in rural Malawi. Am Sociol Rev 74: 171–189.

76. Dunkle KL, Wingood GM, Camp CM, DiClemente RJ (2010) Economically motivated relationships and transactional sex among unmarried African American and white women: results from a U.S. national telephone survey. Public Health Rep 125: Suppl 490–100.

77. Luke N (2006) Exchange and condom use in informal sexual relationships in urban Kenya. Econ Dev Cult Change 54: 319–348.

78. Hill AB (1965) The environment and disease: association or causation? Proc R Soc Med 58: 295–300.

Food Insecurity is a Barrier to Prevention of Mother-to-Child HIV Transmission Services in Zimbabwe: A Cross-Sectional Study

Sandra I. McCoy, Raluca Buzdugan,
Angela Mushavi, Agnes Mahomva,
Frances M. Cowan, and Nancy S. Padian

5.1 BACKGROUND

It is widely recognized that attrition from the prevention of mother-to-child HIV transmission (PMTCT) cascade is a significant obstacle to achieving UNAIDS' and the World Health Organization (WHO)'s goal to eliminate mother-to-child transmission by 2015 [1-4]. The PMTCT cascade is a series of services that HIV-positive pregnant women and their infants need to receive in order to prevent HIV transmission, including antenatal care (ANC), HIV testing, and antiretroviral therapy (ART) or antiretroviral (ARV) prophylaxis [5]. However, 49% of HIV-infected pregnant women in sub-Saharan Africa are lost between ANC registration and delivery and miss some or all essential PMTCT services [6]. Furthermore, high rates of loss to follow-up among women initiating ART

© McCoy et al.; licensee BioMed Central. 2015; BMC Public Health, 2015, 15:420; DOI: 10.1186/s12889-015-1764-8. Distributed under the terms of the Creative Commons Attribution 4.0 International License (http://creativecommons.org/licenses/by/4.0/).

under 'Option B+', [7] WHO's PMTCT strategy whereby all pregnant and breastfeeding women receive lifelong ART, [8] has renewed emphasis on the importance of reducing barriers to uptake of PMTCT services.

In many settings, economic factors are cited as barriers to ANC and PMTCT services, [9,10] including facility-based delivery, [11] initiation of ART and ARV prophylaxis, and retention in HIV care [12]. Here we focus on one aspect of socioeconomic position: food security. People are considered food secure when they have adequate physical, social, and economic access to sufficient, safe and nutritious food that meets their dietary needs and food preferences for an active and healthy life [13]. Food insecurity is increasingly recognized as exacerbating the HIV/AIDS epidemic by increasing engagement in HIV-related risk behaviors, [14-17] and among people living with HIV, undermining ART adherence and retention in care [18-20]. However, few studies have examined the relationship between food insecurity and utilization of PMTCT services and MTCT [19,20].

There are several pathways through which food insecurity might affect women's use of health services and increase vertical HIV transmission [21]. First, food insecurity might result in avoidance or delay of maternal health services because of its overlap with socioeconomic position [22,23] and the real or perceived costs of ANC, facility delivery, and/or HIV prevention and care services. Second, food insecurity is associated with undernutrition, which among HIV-infected women is associated with preterm delivery, low birth weight, and MTCT [24-26]. In addition, food insecure HIV-infected women may be less likely to adhere to ART/ARV prophylaxis, [27] and they may exclusively breastfeed their infants for shorter periods of time, heightening the risk of onward transmission as they resort to mixed feeding [28,29]. Lastly, food insecurity may influence women's receipt of PMTCT services through mental health pathways related to the anxiety and stress associated with real or perceived hunger [30,31]. Stress and depression, in turn, may affect service utilization and obstetric and child health outcomes [32,33]. Together, these compelling pathways support the hypothesis that food insecurity may undermine global efforts to achieve elimination of mother-to-child HIV transmission.

We examined the relationship between food security and PMTCT in Zimbabwe, where a generalized HIV epidemic coexists with food insecurity, hunger, and undernutrition. Thirty-three percent of Zimbabwe's population is undernourished, [34] and the 2013 Global Hunger Index was 16.5, indicating serious levels of hunger [35]. Moreover, 12% of pregnant women are HIV-positive [36]. The objectives of our analyses were to: 1) determine the prevalence of food insecurity among women with a recent birth; 2) explore whether food insecurity is associated with receipt of PMTCT-related services; and 3) examine the association between food insecurity and MTCT.

5.2 METHODS

5.2.1 Study Population

We analyzed data from a 2012 cross-sectional survey of mother/caregiver-infant pairs conducted as part of the impact evaluation of Zimbabwe's Accelerated National PMTCT Program [5,37]. The survey targeted women who were ≥16 years old and biological mothers or caregivers of infants (alive or deceased) born 9–18 months earlier in order to capture MTCT during pregnancy, delivery and breastfeeding. The primary outcomes of the impact evaluation were MTCT and HIV-free infant survival. Because the analyses presented in this paper use data from the 2012 survey, which were the baseline data for the parent impact evaluation, we describe the association between food security and engagement in PMTCT services before the Ministry of Health and Child Care's (MoHCC) implementation of PMTCT strategy 'Option A'. We restricted the sample of 9,018 mothers and caregivers to 8,662 biological mothers and their eligible infants by excluding 356 (3.9%) caregiver/infant pairs.

5.2.2 Sampling Strategy and Data Collection

The two-stage sampling strategy has been previously described [37,38]. Five provinces (Harare, Mashonaland West, Mashonaland Central, Manicaland, and Matabeleland South) were purposefully selected to include Zimbabwe's capital, rural communities with higher and lower HIV prevalence, and both Shona and Ndebele ethnic groups. In the first stage, we randomly selected 157 catchment areas from 699 health facilities offering PMTCT services, proportionate to the number of facilities per district. In the second stage, in each catchment area, a pre-determined proportion of eligible infants was randomly sampled, depending on the size of the catchment area. Potentially eligible infants and their mothers/caregivers were identified by pooling information from: 1) community health workers, 2) immunization registers from both sampled and nearby health facilities, and 3) peer referral. Together, this approach efficiently identified eligible participants without screening all households and captured mother-infant pairs who did not utilize any health services and those who accessed care outside of their area of residence. Mothers providing written informed consent completed an anonymous interviewer-administered survey about maternal and household demographics, health services accessed during pregnancy and after delivery, and behaviors germane to MTCT (e.g., breastfeeding).

5.2.3   Food Security Status

Household food security was determined with a subset of questions from the Household Food Insecurity Access Scale (HFIAS) [30]. Due to interview time constraints, we selected three questions for inclusion, one from each domain of food access of the HFIAS: 1) anxiety and uncertainty about household food supply; 2) insufficient quality, including food variety and preferences; and 3) insufficient food intake and its physical consequences [30,39]. Women were asked how often, in the last 4 weeks, they worried that their household would not have enough food (anxiety/uncertainty), how often they were not able to eat preferred foods because of lack of resources (insufficient quality), and whether anyone in the household went to bed hungry (insufficient intake).

Based on the distribution of these responses, consideration of the recommendations for categorizing responses to the full HFIAS, and examination of other food security scales [40], we determined an algorithm to classify households into three mutually exclusive groups: food secure, moderately food insecure, and severely food insecure. Severe food insecurity was defined as ≥1 household member going to bed hungry (even if infrequently or rarely) or "often" worrying (more than 10 times in the last month) about food access or food quality. Households were classified as having moderate food insecurity if they "sometimes" (3–10 times in the last month) worried about food access or quality. Food secure households experienced either none of the food insecurity conditions or they only rarely worried about food access or quality. We assumed that household food security status in the previous 4 weeks was strongly correlated with what food security status would have been during pregnancy, 9–18 months prior. We excluded seven women without food security information from the analysis.

5.2.4   Assessment of HIV Status

Living mothers and infants provided blood spot samples for HIV testing, which were air-dried onto filter papers and stored at room temperature until biweekly transport to the laboratory. Maternal samples were tested for HIV-1 antibody using AniLabsytems EIA kit (AniLabsystems Ltd, OyToilette 3, FIN-01720, Vantaa, Finland) with positive specimens confirmed using Enzygnost Anti-HIV 1/2 Plus ELISA (Dade Behring, Marburg, Germany) and discrepant results resolved by Western Blot. Samples from HIV-exposed infants and infants of mothers with unavailable samples were tested for HIV with DNA polymerase chain reaction (Roche Amplicor HIV-1 DNA Test, version 1.5). Results were available for 97.8% and 97.2% of women and HIV-exposed infants, respectively. Women were able to receive their HIV test results at the local health facility up to 3 months after the survey using a card with a barcode of their unique identification number.

5.2.5   Data Analyses

We first compared socio-demographic characteristics and service utilization stratified by food security status. We examined the following maternal health services: ANC (any and the WHO-recommended ≥4 visits [41]), gestational age in weeks at ANC registration (WHO recommends the first visit should occur in first trimester [41]), HIV testing during ANC or labor and delivery (or prior knowledge of HIV-positive serostatus), facility-based delivery, and postnatal visit attendance (6–8 weeks postpartum). Among HIV-infected women, we examined reported use of maternal and infant ART/ARV prophylaxis, infant co-trimoxazole prophylaxis, exclusive breastfeeding (≥1 month), and MTCT, stratified by food security status. We also examined a combined category indicating "completion" of the cascade including the following key services: ≥4 ANC visits, HIV testing, facility-based delivery, postnatal visit attendance, and among HIV-infected women, report of maternal and infant ART or ARV prophylaxis and co-trimoxazole prophylaxis. Missing values of PMTCT services were <1%; in those few cases, women were classified as not having received the service.

   We conducted an exploratory analysis to describe the association between food security and completion of the PMTCT cascade and MTCT using Poisson regression models. With cross-sectional data, the exponentiated parameter estimates represent prevalence ratios (PR) [42-44]. The fully adjusted models contain all covariates specified a priori for inclusion (see below) and key services or behaviors not hypothesized to lie on the causal pathway between food insecurity and the outcome. Covariates with variance inflation factors >10 (indicating multicollinearity) were examined for correlation with food security status and if necessary, excluded [45]. We present PRs and 95% confidence intervals (CI) computed with linearized standard errors to account for the sample design.

Several covariates, which likely preceded pregnancy, were considered for inclusion in models as potential confounders: province, mother's age, religion, tribe, being married or having a regular sexual partner, mother's highest educational level, household size, lifetime births, and the building materials of the best building on the homestead. Additionally, we created a household asset index, divided into quartiles, using principal component analysis with a polychoric correlation matrix [46-48]. We also included a variable to indicate the infant's age in months at the time of the survey (or age the infant would have been, if deceased), indicative of the time elapsed between the pregnancy and the interview to account for recall bias. No more than 1% of any covariate was missing. All analyses were conducted with STATA 12 (College Station, Texas) and were weighted to account for the varying sampling fraction by catchment area and 1.1% survey non-response.

5.2.6   Human Subjects Protection

The Medical Research Council of Zimbabwe and the ethical review boards at the University of California, Berkeley and University College London approved this study.

5.3   RESULTS

5.3.1   Participant Characteristics

The weighted population included 8,790 eligible mothers (based on 8,655 observations). The average age of women was 26.7 years, 93% were married or had a regular sexual partner, and they had an average of 2.7 lifetime births (Table 1). Overall, 4,305 (49%) women reported living in food secure households, 2,906 (33%) were living in moderately food insecure households, and 1,578 (18%) were living in severely food insecure households. Women living in moderately or severely food insecure households were less likely to be from Harare, less likely to have a husband or main partner (92% vs. 94%), had less education, fewer assets, and larger household sizes (5.3 vs. 5.0 members). Of the women with HIV test results, 1,075 (12.4%) were HIV-infected; including 9.8%, 12.4%, and 19.4% of women living in food secure, moderately food insecure, and severely food insecure households, respectively (p < 0.01).

TABLE 5.1 Sociodemographic characteristics of participants, Zimbabwe, 2012; Women were ≥16 years old and mothers of infants (alive or deceased) born 9–18 months prior to the interview[a]

Characteristic	Total (N=8,790)	Household food security status[b]		
		Food secure (n=4,305)	Moderate food insecurity (n=2,906)	Severe food insecurity (n=1,578)
	N (%)	N (%)	N (%)	N (%)
Province				
Harare	1,529 (17.4)	930 (21.6)	439 (15.1)	160 (10.1)
Manicaland	3,564 (40.6)	1,449 (33.7)	1,366 (47.0)	749 (47.5)
Mashonaland Central	1,503 (17.1)	818 (19.0)	504 (17.3)	181 (11.5)
Mashonaland West	1,339 (15.2)	713 (16.6)	413 (14.2)	214 (13.6)
Matabeleland South	855 (9.7)	396 (9.2)	184 (6.3)	275 (17.4)
Age, years (mean, SE)	26.7 (0.09)	26.2 (0.12)	27.0 (0.12)	27.4 (0.21)
Married or has a regular sexual partner	8,152 (92.7)	4,028 (93.6)	2,691 (92.6)	1,432 (90.8)
Education, highest completed				
No education	274 (3.1)	99 (2.3)	123 (4.2)	52 (3.3)
Primary school (Standard 7)	2,458 (28.0)	966 (22.4)	854 (29.4)	638 (40.4)
Some secondary school	2,469 (28.1)	1,120 (26.0)	855 (29.4)	494 (31.3)
"O" Level or more (Grade 11)	3,589 (40.8)	2,120 (49.3)	1,073 (36.9)	396 (25.1)
Ethnicity				
Shona	7,341 (83.5)	3,646 (84.7)	2,516 (86.6)	1,179 (74.7)
Ndebele	586 (6.7)	297 (6.9)	116 (4.0)	172 (10.9)
Kalanga/Other	862 (9.8)	361 (8.4)	274 (9.4)	227 (14.4)
Household size (mean, SE)	5.2 (0.05)	5.0 (0.06)	5.2 (0.05)	5.5 (0.08)
Asset Index (quartile)				
1st (lowest)	2,463 (28.0)	862 (20.0)	874 (30.1)	726 (46.0)
2nd	1,624 (18.5)	748 (17.4)	559 (19.2)	317 (20.1)
3rd	2,022 (23.0)	1,016 (23.6)	687 (23.6)	320 (20.3)
4th (highest)	2,681 (30.5)	1,679 (39.0)	786 (27.1)	216 (13.7)

TABLE 5.1   *(Continued)*

Characteristic	Total (N = 8,790)	Household food security status[b]		
		Food secure (n = 4,305)	Moderate food insecurity (n = 2,906)	Severe food insecurity (n = 1,578)
	N (%)	N (%)	N (%)	N (%)
Lifetime births (mean, SE)	2.7 (0.04)	2.4 (0.04)	2.8 (0.04)	3.1 (0.06)
Infant alive	8,726 (99.3)	4,278 (99.4)	2,882 (99.2)	1,565 (99.2)
Infant's age, months (mean, SE)[c]	13.7 (0.04)	13.6 (0.05)	13.7 (0.07)	13.6 (0.08)
Mother HIV-infected	1,075 (12.4)	414 (9.8)	358 (12.4)	304 (19.4)

SE: Linearized standard error.

[a] Weighted counts and proportions presented in the table. Numbers may not sum to column totals due to missing data. Percentages may not add to 100 due to rounding.

[b] Food security determined from a subset of questions from the Household Food Insecurity Access Scale (HFIAS) [30].

[c] Age of infants who were alive as well as the age deceased infants would have been at the time of the survey.

5.3.2   Food Insecurity and Receipt of Maternal Health Services

Food insecurity was inversely associated with use of ANC services, with 95%, 94%, and 92% of women from food secure, moderately food insecure, and severely food insecure households reporting attendance at ≥1 ANC visit, respectively ($p < 0.01$, Table 2), although food insecurity was not associated with ≥4 ANC visits nor the timing of the first ANC visit. Compared to women from food secure households, women from moderately or severely food insecure households were significantly less likely to know their HIV status during pregnancy or labor and delivery, were significantly less likely to deliver in a health facility, and were less likely to report attending the postnatal visit. Overall, completion of all key steps in the PMTCT cascade was reported by 49% of women from food secure households, 45% of women from moderately food insecure households, and 38% of women from severely food insecure households (adjusted $PR_a = 0.95$, 95% CI: 0.90-1.00 (moderate food insecurity vs. food secure), $PR_a = 0.86$, 95% CI: 0.79, 0.94 (severe food insecurity vs. food secure), adjusted Wald test $p < 0.01$, Table 3). There was no association between food insecurity and completion of the cascade when the analysis was restricted to HIV-infected women.

TABLE 5.2 Food security status and receipt of services in the PMTCT cascade, Zimbabwe, 2012[a]

Service	Total (N = 8,790) N (%)	Household food security status[b]		
		Food secure (n = 4,305) N (%)	Moderate food insecurity (n = 2,906) N (%)	Severe food insecurity (n = 1,578) N (%)
Antenatal care (ANC)				
Any	8,287 (94.2)	4,091 (95.0)	2,737 (94.3)	1,450 (92.0)**
≥4 visits	5,627 (64.5)	2,801 (65.6)	1,855 (64.3)	971 (62.2)
Gestational age (months) at booking (mean, SE)	5.1 (0.06)	5.1 (0.08)	5.1 (0.06)	5.0 (0.08)
Tested for HIV infection in ANC or labor and delivery (L&D) or knew HIV-infected	8,117 (92.4)	4,043 (93.9)	2,689 (92.5)	1,385 (87.8)**
Health facility delivery	6,747 (76.8)	3,469 (80.6)	2,199 (75.7)	1,079 (68.4)**
Postnatal visit attendance	8,109 (92.4)	4,040 (93.9)	2,647 (91.3)	1,422 (90.1)**
If HIV-infected (n = 1,075):				
Received ART or ARVs	639 (59.5)	242 (58.6)	215 (60.2)	182 (59.9)
Infant prophylaxis	673 (62.9)	250 (61.1)	227 (63.4)	196 (64.7)
Co-trimoxazole prophylaxis	475 (44.2)	184 (44.6)	166 (46.5)	124 (40.9)
Exclusive breastfeeding (ever)	959 (95.3)	364 (95.9)	314 (92.0)	281 (98.3)**
Infant infected	93 (9.0)	33 (8.2)	22 (6.2)	39 (13.3)**
Received all key maternal health services[c]	4,027 (45.8)	2,121 (49.3)	1,301 (44.8)	605 (38.3)**
HIV-infected women	299 (27.8)	111 (26.7)	107 (30.0)	82 (26.8)
HIV-uninfected women	3,727 (49.1)	2,011 (52.8)	1,194 (47.4)	523 (41.5)**

SE: Linearized standard error.

Design-based chi-squared p-value: *p < 0.05, **p < 0.01.

[a] Weighted counts and proportions presented in the table. Numbers may not sum to column totals due to missing data. Percentages may not add to 100 due to rounding.

[b] Food security determined from the responses to a subset of questions from the Household Food Insecurity Access Scale (HFIAS) [30].

[c] Completed at least 4 ANC visits, was tested for HIV infection or already knew HIV-positive serostatus, delivered infant in a health facility, and attended the postnatal visit. Among HIV-infected women, must also have reported maternal and infant ART or ARV prophylaxis and receipt of infant co-trimoxazole prophylaxis.

TABLE 5.3   Association between household food security and completion of services in the PMTCT cascade and MTCT, Zimbabwe, 2012

Household food security status	All women: completion of key PMTCT services[a]				HIV-infected women: MTCT[b]			
	Unadjusted		Adjusted		Unadjusted		Adjusted	
	PR	95%CI	PR	95%CI	PR	95%CI	PR	95%CI
Food secure	1	—	1	—	1	—	1	—
Moderate food insecurity	0.91	(0.86, 0.96)**	0.95	(0.90, 1.01)	0.75	(0.46, 1.24)	0.68	(0.43, 1.08)
Severe food insecurity	0.78	(0.70, 0.86)**	0.86	(0.79, 0.94)**	1.62	(1.04, 2.52)*	1.42	(0.89, 2.26)

PR: prevalence ratio; CI: confidence interval.

P-value: $*p < 0.05$, $**p < 0.01$.

[a]Regression model of a weighted sample of 8,655 women. Outcome: combined variable indicating at least 4 ANC visits, tested for HIV infection or already knew HIV-positive serostatus, delivered infant in a health facility, attended the postnatal visit, and among HIV-infected women, report of maternal and infant ART or ARV prophylaxis and receipt of infant co-trimoxazole prophylaxis. Adjusted model includes province, maternal age, whether the woman has a husband or regular partner, education, tribe, religion, household size, building material of the best structure, an asset index created using principal component analysis, and whether the infant was alive at the time of the survey.

[b]Regression model of a weighted sample of 1,058 HIV-infected women who had infants with HIV test results. Adjusted model includes province, maternal age, whether the woman has a husband or regular partner, education, tribe, religion, household size, building material of the best structure, an asset index created using principal component analysis, and whether the infant was alive at the time of the survey.

5.3.3   Food Insecurity and MTCT

Food insecurity was not associated with maternal or infant receipt of ART/ARV prophylaxis. However, women living in severely food insecure households were the most likely to have ever exclusively breastfed their infant (98%, vs. 96% of women in food secure and 92% of women in moderately food insecure households, p < 0.01). Of HIV-exposed infants, 13.3% of infants from severely food insecure households were HIV-infected compared to 8.2% of infants from food secure households (PR = 1.62, 95% CI: 1.04, 2.52, Tables 2 & 3). After adjustment for covariates, this association was attenuated (PR$_a$ = 1.42, 95% CI: 0.89, 2.26). There was no increased likelihood of MTCT among women from moderately food insecure households in unadjusted or adjusted analyses (PR$_a$ = 0.68, 95% CI: 0.43, 1.08).

5.4   DISCUSSION

In this analysis of women with a recent birth in Zimbabwe, we found that more than half reported living in moderate or severely food insecure households in the month prior to the survey. Compared to women from food secure households, women from food insecure households were more likely to be HIV-infected. Consistent with previous qualitative studies, [49] we found that food insecurity may be an important barrier to uptake of some PMTCT services: in unadjusted analyses, food insecurity was inversely associated with ANC, knowing one's HIV status, facility-based delivery, and postnatal visit attendance. When services were examined together, and after adjustment for covariates, women who reported severe food insecurity were 14% less likely to complete all recommended maternal and infant health services for PMTCT compared to food secure women. Although the effect sizes are modest and absolute differences are small, these findings suggest that among a subgroup of pregnant women, severe food insecurity is an important barrier to some maternal health services.

Among HIV-infected women, we unexpectedly found that there was no association between food insecurity and completion of the PMTCT cascade. This might be due to the small sample size, women's motivation to protect their infant from HIV infection, or nutritional support provided during pregnancy and postpartum that partially mitigates household food insecurity. However, women from severely food insecure households were 42% more likely to have an HIV-infected infant compared to women from food secure households, although this finding was not statistically significant after adjustment for covariates. Although food security's association with vertical transmission has been speculated, [21,50] this analysis is the first, to our knowledge, to provide empirical data supporting this hypothesis. We must nevertheless interpret these findings with caution because a strong 'dose-response' relationship between food insecurity and MTCT was not observed, as women in moderately food insecure households had the lowest proportions of exclusive breastfeeding and MTCT.

Although we found that food insecurity was associated with ever attending ANC, we found no association between food insecurity and attending the WHO-recommended ≥4 ANC visits or the timing of ANC registration. There are several potential explanations for this finding. In facilities that charge a fee, the fees often cover the bundle of services from ANC booking through the 6-week postnatal visit. Thus, once engaged in ANC, women may be retained in the cascade and continue to receive ANC care and deliver at the health facility. Women who don't attend ANC may therefore also be more likely to deliver at home and miss the postnatal visit. Although this study found only small but statistically significant reductions in utilization of each service among women who reported being moderately or severely food insecure, losses from the cascade are additive, as was demonstrated by a cohort study in Cameroon, Côte d'Ivoire, South Africa, and Zambia, where unremarkable levels of attrition of HIV-infected mothers from individual steps in the cascade coupled with poor adherence resulted in 49% of HIV-exposed infants not being protected by a prophylactic ARV regimen [51].

An unanswered question is the pathway(s) through which food insecurity might impede service utilization. It is possible that food insecurity in this analysis is simply a proxy for poverty, and our findings are reflective of the difficult choices food insecure women must make between food (and other goods and services) and the costs associated with health care, including transport and fees. Certainly, food insecurity is highly correlated with socioeconomic position [22,23] and has been shown to be the strongest measure of socio-economic position associated with HIV and HSV-2 risk among young women in Zimbabwe, potentially due to engagement in risk behavior to obtain food or other essential goods and services [52]. This might be the most likely explanation for the inverse association we identified between food insecurity and ANC and other services that require or are perceived to require payment. Although Zimbabwe is moving toward elimination of user fees for maternal and child health services, some facilities still charge a fee that may be cost-prohibitive.

However, this economic explanation does not fully explain the finding that severe food insecurity was not associated with receipt of maternal and infant

ART/ARVs, but may nevertheless be associated with MTCT, although the width of the confidence limits suggests substantial uncertainty in this association. One explanation might be that although both food secure and insecure women and infants were equally likely to receive ART/ARV prophylaxis, women who were food insecure were less likely to adhere to treatment. Food insecurity is known to reduce ART adherence in non-pregnant populations, [27] although this study is not able to test this hypothesis. Another possible explanation is more complex: severe food insecurity increases MTCT risk due to cumulative loss of women from the cascade coupled with an increased risk of MTCT associated with undernutrition [25,26] and the possible increased propensity for mixed feeding [28]. This study was unable to explore more complex patterns of breast-feeding (such as duration of exclusive breastfeeding) to support or refute this hypothesis, and our simple measure of exclusive breastfeeding (ever) found that nearly all women exclusively breastfed at some point. Furthermore, this second-ary data analysis used cross-sectional data from a study that was not designed to specifically examine this question, so we do not have prospective data nor infor-mation on several causal intermediates, including nutritional indicators such as women's body mass index (BMI) and micronutrient status, to examine these pathways. Nonetheless, these data provide both compelling empirical evidence about these hypothesized relationships and also reveal important gaps for future research.

An important issue when considering these findings is the measurement of food security [53,54]. In this study, household food security was measured in the 4 weeks prior to the survey, a common reference period used by other scales [30,55,56] in order to balance the tradeoffs between recall bias (which favors a shorter recall period) and 'telescoping errors' (a phenomenon associ-ated with short recall periods whereby events outside of the recall window are erroneously reported) [57,58]. We made the essential temporality assumption that a household's recent food security status was highly correlated with its food security status during the woman's pregnancy, 9–18 months prior. The valid-ity of this assumption is unknown; it depends on how much household food access changed over time and season, which was not measured in the survey. Nevertheless, this issue could be overcome in future prospective studies by conducting a simple food security assessment at ANC registration to identify food insecure women who are at risk of both undernutrition and of missing key PMTCT services.

Another key measurement issue is that food security was measured using only part of the HFIAS, a validated scale, which adds some uncertainty to the classification of food security. Nevertheless, our inclusion of a question from each dimension of food access [30] in addition to the correlation between our parameterization of food security and other dimensions of household socio-economic position increases confidence in our classification scheme. A strength of our measurement of food security is that women themselves reported household food security status, the individuals who are typically responsible for a household's food supply and meal preparation in Zimbabwe. This is also important because food insecurity at the individual level may be prevalent even in wealthier households due to unequal intra-household allocation of food, which, for example, can result in women eating last or having less access to fats, protein, or micronutrient-rich foods [59-61]. Nevertheless, as with all measures of food insecurity, food security may be subject to underreporting because of its sensitive nature [62].

Our analysis has other important limitations. We used cross-sectional data and therefore cannot make inferences about causation. Further, women self-reported receipt of healthcare services. Moreover, although our data are representative of the communities from which the sample was selected, they are not representative of all regions in Zimbabwe, and it is possible that the relationship between food insecurity and service utilization are different in other parts of the country. In addition, although our strategy to create a sampling frame of 9–18 month old infants in the community was comprehensive, it is possible that some mother-infant pairs were missed. Lastly, women's and infant's HIV status was measured at the time of the survey, 9–18 months postpartum. Although we have assumed that women who were HIV-infected at the time of the survey were also HIV-infected during their pregnancy, it is possible that a small proportion were infected during pregnancy or postpartum. Likewise, infants who were still breastfeeding at the time of the survey remained at risk of MTCT, so we may not have captured all possible infant infections.

5.5 CONCLUSIONS

This analysis is the first to our knowledge to examine the association between food insecurity and receipt of services in the PMTCT cascade. Our findings suggest that severe food insecurity may impede the receipt of some services among pregnant and postpartum women and may influence MTCT; these relationships will need to be confirmed by other prospective observational studies. In addition, the growing body of studies that examine the effect of food and cash transfers as a mechanism to mitigate food insecurity and improve HIV-related outcomes [63,64] will also contribute to understanding these relationships, including whether food assistance programs have unexpected spillover benefits on maternal and child health. Regardless of its effect on MTCT, the high prevalence of food insecurity among women with a recent birth in Zimbabwe provides additional support for integrated food and nutrition programs for pregnant women.

REFERENCES

1. World Health Organization, UNICEF, Interagency Task Team of Prevention of HIV Infection in Pregnant Women M, and their Children. Guidance on Global Scale-Up of the Prevention of Mother to Child Transmission of HIV. Geneva: WHO; 2007.
2. UNAIDS. Report on the Global AIDS Epidemic. Geneva; 2012.
3. Joint United Nations Programme on HIV/AIDS (UNAIDS): Global Plan Towards the Elimination of New HIV Infections Among Children By 2015 and Keeping their Mothers Alive, 2011–2015. Geneva; 2011.
4. Mahy M, Stover J, Kiragu K, Hayashi C, Akwara P, Luo C, et al. What will it take to achieve virtual elimination of mother-to-child transmission of HIV? An assessment of current progress and future needs. Sex Transm Infect. 2010;86 Suppl 2:ii48–55.
5. World Health Organization. Antiretroviral Drugs for Treating Pregnant Women and Preventing HIV Infection in Infants: Recommendations for a public health approach. Geneva; 2010.
6. Sibanda EL, Weller IV, Hakim JG, Cowan FM. The magnitude of loss to follow-up of HIV-exposed infants along the prevention of mother-to-child HIV transmission continuum of care: a systematic review and meta-analysis. AIDS. 2013;27(17):2787–97.
7. Tenthani L, Haas AD, Tweya H, Jahn A, van Oosterhout JJ, Chimbwandira F, et al. Retention in care under universal antiretroviral therapy for HIV-infected pregnant and breastfeeding women ('Option B+') in Malawi. AIDS. 2014;28(4):589–98.
8. World Health Organization. Programmatic Update. Use of Antiretroviral Drugs for Treating Pregnant Women And Preventing HIV Infection in Infants. In. Geneva; 2012.
9. HIarlaithe MO, Grede N, de Pee S, Bloem M. Economic and Social Factors are Some of the Most Common Barriers Preventing Women from Accessing Maternal and Newborn Child Health (MNCH) and Prevention of Mother-to-Child Transmission (PMTCT) Services: A Literature Review. AIDS Behav. 2014;18 Suppl 5:516–30.
10. Clouse K, Schwartz S, Van Rie A, Bassett J, Yende N, Pettifor A. "What they wanted was to give birth; nothing else": barriers to retention in option B+ HIV care among postpartum women in South Africa. J Acquir Immune Defic Syndr. 2014;67(1):e12–8.
11. Moyer CA, Mustafa A. Drivers and deterrents of facility delivery in sub-Saharan Africa: a systematic review. Reprod Health. 2013;10:40.
12. Gourlay A, Birdthistle I, Mburu G, Iorpenda K, Wringe A. Barriers and facilitating factors to the uptake of antiretroviral drugs for prevention of mother-to-child transmission of HIV in sub-Saharan Africa: a systematic review. J Int AIDS Soc. 2013;16(1):18588.

13. Food and Agriculture Organization of the United Nations. The State of Food Insecurity in the World. Rome; 2010.

14. Tsai AC, Hung KJ, Weiser SD. Is food insecurity associated with HIV risk? Cross-sectional evidence from sexually active women in Brazil. PLoS Med. 2012;9(4):e1001203.

15. Miller CL, Bangsberg DR, Tuller DM, Senkungu J, Kawuma A, Frongillo EA, et al. Food Insecurity and Sexual Risk in an HIV Endemic Community in Uganda. AIDS Behav. 2011;15(7):1512–9.

16. Oyefara JL. Food insecurity, HIV/AIDS pandemic and sexual behaviour of female commercial sex workers in Lagos metropolis, Nigeria. Sahara J. 2007;4(2):626–35.

17. Weiser SD, Leiter K, Bangsberg DR, Butler LM, Percy-de Korte F, Hlanze Z, et al. Food insufficiency is associated with high-risk sexual behavior among women in Botswana and Swaziland. PLoS Med. 2007;4(10):1589–97. discussion 1598.

18. Weiser SD, Fernandes KA, Brandson EK, Lima VD, Anema A, Bangsberg DR, et al. The association between food insecurity and mortality among HIV-infected individuals on HAART. J Acquir Immune Defic Syndr. 2009;52(3):342–9.

19. Weiser SD, Hatcher A, Frongillo EA, Guzman D, Riley ED, Bangsberg DR, Kushel MB. Food Insecurity Is Associated with Greater Acute Care Utilization among HIV-Infected Homeless and Marginally Housed Individuals in San Francisco. Journal of general internal medicine 2012.

20. Weiser SD, Tsai AC, Gupta R, Frongillo EA, Kawuma A, Senkungu J, et al. Food insecurity is associated with morbidity and patterns of healthcare utilization among HIV-infected individuals in a resource-poor setting. AIDS. 2012;26(1):67–75.

21. Weiser SD, Young SL, Cohen CR, Kushel MB, Tsai AC, Tien PC, et al. Conceptual framework for understanding the bidirectional links between food insecurity and HIV/AIDS. Am J Clin Nutr. 2011;94(6):1729S–39.

22. Alaimo K, Briefel RR, Frongillo Jr EA, Olson CM. Food insufficiency exists in the United States: results from the third National Health and Nutrition Examination Survey (NHANES III). Am J Public Health. 1998;88(3):419–26.

23. Barrett CB. Measuring food insecurity. Science. 2010;327(5967):825–8.

24. Young S, Murray K, Mwesigwa J, Natureeba P, Osterbauer B, Achan J, et al. Maternal nutritional status predicts adverse birth outcomes among HIV-infected rural Ugandan women receiving combination antiretroviral therapy. PLoS ONE. 2012;7(8):e41934.

25. Zijenah LS, Moulton LH, Iliff P, Nathoo K, Munjoma MW, Mutasa K, et al. Timing of mother-to-child transmission of HIV-1 and infant mortality in the first 6 months of life in Harare, Zimbabwe. AIDS. 2004;18(2):273–80.

26. Villamor E, Saathoff E, Msamanga G, O'Brien ME, Manji K, Fawzi WW. Wasting during pregnancy increases the risk of mother-to-child HIV-1 transmission. J Acquir Immune Defic Syndr. 2005;38(5):622–6.

27. Singer AW, Weiser SD, McCoy SI. Does Food Insecurity Undermine Adherence to Antiretroviral Therapy? A Systematic Review. AIDS Behav. 2014 Aug 6.

28. Young SL, Plenty AH, Luwedde FA, Natamba BK, Natureeba P, Achan J, et al. Household food insecurity, maternal nutritional status, and infant feeding practices among HIV-infected Ugandan women receiving combination antiretroviral therapy. Matern Child Health J. 2014;18(9):2044–53.

29. Levy JM, Webb AL, Sellen DW. "On our own, we can't manage": experiences with infant feeding recommendations among Malawian mothers living with HIV. Int Breastfeed J. 2010;5:15.

30. Coates J, Swindale A, Bilinsky P. Household Food Insecurity Access Scale (HFIAS) for Measurement of Food Access: Indicator Guide. In. Edited by (FANTA) FaNTAP. Washington, D.C.: United States Agency for International Development; 2007.

31. Garcia J, Hromi-Fiedler A, Mazur RE, Marquis G, Sellen D, Lartey A, et al. Persistent household food insecurity, HIV, and maternal stress in peri-urban Ghana. BMC Public Health. 2013;13:215.

32. Witt WP, Litzelman K, Cheng ER, Wakeel F, Barker ES. Measuring stress before and during pregnancy: a review of population-based studies of obstetric outcomes. Matern Child Health J. 2014;18(1):52–63.

33. Alder J, Fink N, Bitzer J, Hosli I, Holzgreve W. Depression and anxiety during pregnancy: a risk factor for obstetric, fetal and neonatal outcome? A critical review of the literature. J Matern Fetal Neonatal Med. 2007;20(3):189–209.

34. Food and Agriculture Organization of the United Nations (FAO), World Food Programme (WFP), IFAD. The State of Food Insecurity in the World 2012. Economic Growth is necessary but not sufficient to accelerate reduction of hunger and malnutrition. In. Rome: FAO; 2012.

35. von Grebmer K, Ringler C, Rosegrant MW, Olofinbiyi T, Weismann D, Fritschel H, Badiane O, Torero M, Yohannes Y, Thompton J et al. Global Hunger Index. The Challenge of Hunger: Ensuring sustainable food security under land, water and energy stresses. In. Bonn, Washington D. C., Dublin: International Food Policy Research Institute, Concern Worldwide, Welthungerhilfe; 2012.

36. Zimbabwe National Statistics Agency (ZIMSTAT) and ICF International. Zimbabwe Demographic and Health Survey 2010–2011. In. Calverton, Maryland: ZIMSTAT and ICF International, Inc.; 2012.

37. Buzdugan R, McCoy SI, Petersen M, Guay LA, Mushavi A, Mahomva A, Hakobyan A, Cowan FM, Padian N. Feasibility of population-based cross-sectional surveys for estimating vertical HIV transmission: data from Zimbabwe. In: 7th IAS Conference on HIV Pathogenesis, Treatment and Prevention: 2013; Kuala Lumpur; 2013.

38. McCoy SI, Buzdugan R, Ralph LJ, Mushavi A, Mahomva A, Hakobyan A, et al. Unmet Need for Family Planning, Contraceptive Failure, and Unintended Pregnancy among HIV-Infected and HIV-Uninfected Women in Zimbabwe. PLoS ONE. 2014;9(8):e105320.

39. Coates J. Experience and Expression of Food Insecurity Across Cultures: Practical Implications for Valid Measurement. In. Washington, D.C.: Food and Nutrition Technical Assistance Project, Academy for Educational Development; 2004.

40. Bickel G, Nord M, Price C, Hamilton W, Cook J. Guide to Measuring Household Food Security. In: Measuring Food Security in the United States: Reports of the Federal Interagency Food Security Measurement Project. Alexandria, VA: United States Department of Agriculture; 2000.

41. World Health Organization. Antenatal Care Randomized Trial: Manual for the Implementation of the New Model. Geneva; 2002.

42. Zocchetti C, Consonni D, Bertazzi PA. Relationship between prevalence rate ratios and odds ratios in cross-sectional studies. Int J Epidemiol. 1997;26(1):220–3.

43. Greenland S. Interpretation and choice of effect measures in epidemiologic analyses. Am J Epidemiol. 1987;125(5):761–8.

44. Barros AJ, Hirakata VN. Alternatives for logistic regression in cross-sectional studies: an empirical comparison of models that directly estimate the prevalence ratio. BMC Med Res Methodol. 2003;3:21.

45. Kutner MH, Nachtsheim CJ, Neter J, Li W. Applied Linear Statistical Models. 5th ed. New York, New York: McGraw-Hill/Irwin; 2005.

46. Kolenikov S, Angeles G. The Use of Discrete Data in Principal Component Analysis With Applications to Socio-Economic Indices. In: CPC/MEASURE Working paper. Chapel Hill, North Carolina: MEASURE Evaluation; 2004.

47. Vyas S, Kumaranayake L. Constructing socio-economic status indices: how to use principal components analysis. Health Policy Plan. 2006;21(6):459–68.

48. Filmer D, Pritchett L. Estimating Wealth Effect Without Expenditure Data—Or Tears: An Application to Educational Enrollments in States of India. Demography. 2001;38:115–32.

49. Iroezi ND, Mindry D, Kawale P, Chikowi G, Jansen PA, Hoffman RM. A qualitative analysis of the barriers and facilitators to receiving care in a prevention of mother-to-child program in Nkhoma, Malawi. Afr J Reprod Health. 2013;17(4 Spec No):118–29.

50. Anema A, Vogenthaler N, Frongillo EA, Kadiyala S, Weiser SD. Food insecurity and HIV/AIDS: current knowledge, gaps, and research priorities. Curr HIV/AIDS Rep. 2009;6(4):224–31.

51. Stringer EM, Ekouevi DK, Coetzee D, Tih PM, Creek TL, Stinson K, et al. Coverage of nevirapine-based services to prevent mother-to-child HIV transmission in 4 African countries. JAMA. 2010;304(3):293–302.

52. Pascoe SJS, Hargreaves J, Langhaug L, Webster M, Jaffar S, Hayes R, Cowan FM. Poverty, food insufficiency and HIV infection and sexual behaviour among young rural Zimbabwean women. PLoS One 2014, In press.

53. Anema A, Fielden SJ, Castleman T, Grede N, Heap A, Bloem M. Food Security in the Context of HIV: Towards Harmonized Definitions and Indicators. AIDS Behav. 2014;18 Suppl 5:476–89.

54. Maxwell D, Coates J, Vaitla B. How Do Different Indicators of Household Food Security Compare? Empirical Evidence from Tigray. In. Edited by Center FI. Somerville, MA: Tufts University; 2013.

55. Deitchler M, Ballard T, Swindale A, Coates J. Introducing a Simple Measure of Household Hunger for Cross-Cultural Use. In: Technical Note No 12. Washington, D.C: Food and Nutrition Technical Assistance II Project, AED; 2011.

56. Jones AD, Ngure FM, Pelto G, Young SL. What are we assessing when we measure food security? A compendium and review of current metrics. Adv Nutr. 2013;4(5):481–505.

57. Deaton A, Grosh M. Designing Household Survey Questionnaires for Developing Countries: Lessons from 15 years of the Living Standards Measurement Study, vol. 1: The World Bank; 2000.

58. Smith LC, Alderman H, Aduayom D. Food Insecurity in Sub-Saharan Africa: New Estimates from Household Expenditure Surveys. In: Research Report 146. Washington, D.C: International Food Policy Research Institute (IFPRI); 2006.

59. Messer E. Intra-household allocation of food and health care: current findings and understandings - introduction. Soc Sci Med. 1997;44(11):1675–84.

60. Luo W, Zhai F, Jin S, Ge K. Intrahousehold food distribution: a case study of eight provinces in China. Asia Pac J Clin Nutr. 2001;10(Suppl):S19–28.

61. Gittelsohn J, Vastine AE. Sociocultural and household factors impacting on the selection, allocation and consumption of animal source foods: current knowledge and application. J Nutr. 2003;133(11 Suppl 2):4036S–41.

62. Pascoe SJ, Hargreaves JR, Langhaug LF, Hayes RJ. Cowan FM: 'How poor are you?' – a comparison of four questionnaire delivery modes for assessing socio-economic position in rural zimbabwe. PLoS ONE. 2013;8(9):e74977.
63. Cantrell RA, Sinkala M, Megazinni K, Lawson-Marriott S, Washington S, Chi BH, et al. A pilot study of food supplementation to improve adherence to antiretroviral therapy among food-insecure adults in Lusaka, Zambia. J Acquir Immune Defic Syndr. 2008;49(2):190–5.
64. Ivers LC, Chang Y, Gregory Jerome J, Freedberg KA. Food assistance is associated with improved body mass index, food security and attendance at clinic in an HIV program in central Haiti: a prospective observational cohort study. AIDS Res Ther. 2010;7:33.

CHAPTER 6

A Pre-Post Pilot Study of Peer Nutritional Counseling and Food Insecurity and Nutritional Outcomes among Antiretroviral Therapy Patients in Honduras

Kathryn P. Derose, Melissa Felician, Bing Han, Kartika Palar, Blanca Ramírez, Hugo Farías, and Homero Martínez

6.1 BACKGROUND

Food insecurity, defined as "the limited or uncertain availability of nutritionally adequate, safe foods or the inability to acquire personally acceptable foods in socially acceptable ways" [1], and its resultant adverse effects on nutritional status represent important negative influences on HIV outcomes in low-resource settings [2–5]. Studies have found that food insecurity is associated with poor adherence to anti-retroviral therapy (ART) in both resource-poor regions [4, 6–8], as well as among vulnerable populations in resource-rich settings [9–12]. Because of the link among food insecurity, poor nutritional status, and adverse HIV outcomes, the World Health Organization (WHO) recommends that interventions to promote initiation of and adherence to ART include attention to a sufficient and balanced diet [13]. Research as to how to provide a healthy diet among PLHIV has identified several approaches: nutrition supplementation (e.g., specialized foods) with or without nutritional education/counseling; safety nets such as food, cash transfer or vouchers; and livelihood interventions such as small scale agriculture, livestock or sewing [14]. These interventions—especially those providing food assistance—have been found to promote positive effects on nutrition status, quality of life, retention in care, adherence to treatment, and household food security among PLHIV [14, 15]. However, recent research has highlighted that nutrition education and counseling remain weak components of nutritional interventions for PLHIV [14].

Integrating nutrition education and counseling into ART treatment programs is constrained by human capital considerations. Most reported studies of nutritional counseling interventions for PLHIV have relied on professional staff [16–20] and this type of support is limited in low-resource settings. Trained community health workers represent a possible solution. Not only can such an approach contain the cost of delivering nutritional support, but it also offers the linguistic, cultural, and community-building skills to establish rapport with PLHIV. There has been some research into this means of supporting PLHIV: e.g., peer health workers in several countries have been found to be effective in reducing stigma, improving retention in care, and improving quality and outcomes of HIV care [21–26]. However, we are unaware of any reports of the effectiveness of *peer nutritional* counselors among PLHIV.

Here we examine the feasibility and preliminary effectiveness of a nutritional counseling intervention for PLHIV that was adapted for and delivered by peer counselors in Honduras. Specifically, we examined changes between pre- and post-intervention measures of nutritional knowledge, food insecurity, and dietary quality among people receiving ART. Anthropometric measures were also taken to examine any changes pre- and post-intervention to nutritional status, although funding limitations necessitated a short follow-up period, and thus made these secondary outcomes.

6.2 METHODS

6.2.1 Contextual Background

The study took place within an on-going collaboration between RAND (a non-profit research organization), the United Nations World Food Program

(WFP) in Latin America, the Honduran Ministry of Health, and the Honduran National Association of People Living with HIV/AIDS [or *Asociación Nacional de Personas Viviendo con VIH/SIDA en Honduras* (ASONAPVSIDAH)]. This collaboration was formed to evaluate the role that food aid and nutrition counseling played in supporting HIV care retention, uptake of ART and adherence, and selected clinical outcomes [16]. Specifically, an NIMH-funded parent study used a cluster randomized controlled trial design to compare the effects of a monthly household food basket plus nutrition education versus nutrition education alone on ART adherence of HIV patients at four Comprehensive Treatment Centers in Honduras (referred to here as "HIV clinics"). Nutrition education in the parent study was provided by professional nutritionists who were trained in HIV-related nutritional issues by the study investigators. Although political will existed to extend nutritional counseling at all 38 of the nation's HIV clinics, limited financial and human capital resources made this goal difficult to attain. For example, there was no nutrition training program in the country when the parent study began. Thus, a supplemental pilot study was designed to train peer counselors to provide the nutritional counseling, enabling the scale up of counseling across a larger number of the country's HIV clinics. Although the pilot study was modest in scope (pre- and post- study design with no control group and no randomization, limited measures collected and brief follow-up period), we believe it can make an important contribution given the lack of such interventions in the literature and the need for scalable solutions.

6.2.2   Ethics Statement

The study was approved by RAND's Human Subjects Protection Committee and the Institutional Review Board at the National Autonomous University of Honduras. Written informed consent was obtained from all participants.

6.2.3   Study Design

Fourteen government-run HIV clinics were chosen for the counseling extension project, as these represented the most HIV-affected areas in the country. The clinics provided care to nearly 80 % of patients receiving ART in Honduras, and were in geographical proximity to the four study clinics in the parent study, which served as training centers. One or two peer counselors from each center were trained in how to administer the counseling intervention as well as study

assessments (anthropometry, nutritional knowledge, dietary intake, food insecurity). All patients receiving ART at each of the 14 clinics over the course of
one month were offered the peer nutrition counseling (i.e., there was no randomization); those who provided informed consent to participate in the study
were assessed at baseline and 2 months post-baseline, with counseling offered
monthly during the study period. This brief follow-up period was necessitated
by funding limitations and the parent grant's timeline, though was also considered to be sufficient time to observe change in the more proximate outcomes
(e.g., nutritional knowledge and dietary behaviors, including frequency of meals,
and dietary diversity).

6.2.4   Theoretical Background

Our nutritional counseling intervention is based on the information-motivation-behavioral skills model [27, 28]. Specifically, we expected that culturally,
locally, and HIV appropriate nutrition education would improve nutritional
knowledge and dietary intake, which in turn would support a healthy nutritional
status and minimize side effects of HIV and ART, such as nausea, diarrhea, oral
ulcers, etc. We also expected that locally tailored nutritional education could
reduce household food insecurity through improved diet quality, the ability to
make the most of existing food resources, and decreased anxiety about procur-
ing healthy food [29, 30]. Finally, we expected that the nutritional counseling
would be reinforced and made even more salient when delivered by peer counselors, who are in similar circumstances (i.e., living with HIV and on ART) and
can relate to participants' experiences.

6.2.5   Nutritional Counseling Intervention

A nutrition education curriculum with supportive visual aids and a reference
technical manual were developed for use by professional nutritionists in the
parent study after extensive formative research on locally available foods and
dietary patterns among people living with HIV [16]. For the peer counselors, a
simplified version of the training manual was developed and supported by educational materials such as flip charts, brochures, and pocket-sized cards meant
to reinforce session content. The nutrition education curriculum was based on
the WHO/FAO Manual on Nutritional Care and Support for People Living
with HIV/AIDS [13, 31] and our formative research, and covered 5 basic topics

using concepts and language familiar to this population: 1) consuming a balanced diet; 2) food groups; 3) increasing vitamin and mineral intake through a varied diet; 4) food safety; and 5) how to deal with co-morbidities (diarrhea, acute respiratory infection, nausea, mouth sores, loss of appetite). For example, one way that the curriculum translated complex nutritional information into more understandable constructs is reflected in the conceptualization of different "food groups": 1) *alimentos constructores* (building foods, or protein); 2) *alimentos energéticos* (energy foods or carbohydrates and fat); and 3) *alimentos reguladores* (regulating foods or vitamins and minerals) Further, given the setting and the generally low-income population, the curriculum promoted the most affordable, locally available food and culturally appealing ways to prepare them.

Peer counselors were selected from each of the communities where the HIV clinics were located in collaboration with the Ministry of Health and ASONAPVSIDAH, using the following criteria: HIV status (all peer counselors were HIV positive, ART patients); age 18+ years; completion of middle-school; full time availability for the training and project activities, and demonstrated leadership capability. Over a two-month period, peer counselors received 290 h of training (intensive workshop and practical training in one of the parent grant's 4 study clinics—see Table 1 for training topics). Thirty-seven lay workers participated in the workshop training; 20 scored at least 75 % correct during the post-workshop assessments and were invited to participate in the practical training under supervision of the parent grant nutritionists; of these 17 were approved as certified Peer Nutritional Counselors, based on their performance during training. Certification was issued by the Honduran Ministry of Health, WFP and RAND, and was meant to help these workers be considered for future jobs once the project was completed. The peer nutritional counselors were paid for all their time on this project, including training, practicum, and implementation of the pilot.

Each peer nutritional counselor worked out of their local HIV clinic, in a private working space. Participants were recruited through referrals from medical and nursing staff and from the clinic waiting room. Once informed consent was granted, the counselor collected baseline information (survey and anthropometric measurements), delivered the nutritional counseling, and scheduled a date for a follow-up visit in one month, coinciding with the regular monthly visit to the clinic required to retrieve HIV medications. The nutritional counseling, which usually lasted an hour, was tailored to each individual, was highly participatory, and encouraged participants to share concerns about eating and learn about where to obtain certain foods and how to prepare them. The peer counselor shared his/her own experience about how eating a better diet had helped him/her have a better quality of life.

TABLE 6.1   Overview of training provided to peer nutrition counselors

Training Settings	Total Hours	Topics covered
5 day intensive workshop	50	1 Nutrition and healthy eating
		2 HIV, nutrition and food security
		3 ART and adherence
		4 How to manage secondary effects of ART through diet
		5 Anthropometry and dietary assessment
Practical training in "Centro de Atención Integral"	240	1 How to deliver nutritional counseling
		2 How to use supportive educational materials
		3 How to administer evaluation questionnaires (food frequency recall, anthropometry)

6.2.6   Outcomes

Food insecurity was assessed using the validated Latin American and Caribbean Food Security Scale (ELCSA), which captures experiences of household food security over a specified time period (we used previous month for this pilot study), including food quantity and sufficiency, food quality and safety, and anxiety about food supplies [32]. Respondents with children <18 in the household are asked the full 15-item scale; respondents without children are asked the 8-item version of the scale. All questions received "yes" or "no" answers; raw scores were then tabulated as the sum of affirmative answers, with higher scores indicating higher levels of food insecurity. To maximize data in this intervention pilot, we calculated one food insecurity score (on a scale of 0–8) for all respondents and a separate food insecurity score (on a scale of 0–15) for those with children.

Nutritional knowledge was assessed through 13 "true/false" statements that were based on key concepts from the nutrition education curriculum such as the importance of eating a variety of food groups, foods that address anemia, and foods that prevent (and address) constipation and diarrhea. Correctly answered items summed to create a score from 0 (no items answered correctly) to 13 (all items answered correctly).

Dietary intake was assessed using a 24 h qualitative recall. Given the focus of the nutrition education curriculum on increasing the frequency of meals and diversity of food types consumed, supervising nutritionists coded the dietary data for each participant to indicate: 1) the number of times participants reported having eaten in the last 24 h; and 2) the number of different food types consumed at each meal (*energéticos* or carbohydrates and fat, *constructores* or protein, and *reguladores* or vitamins and minerals) to create a measure of dietary diversity.

Nutritional status was assessed using Body Mass Index (BMI) and mid-arm and waist circumference measurements taken by the peer counselors, who were trained by the parent grant nutritionists and standardized according to accepted methods [33]. For descriptive statistics, we use World Health Organization classifications of underweight (<18.5), normal (18.5 to 24.9), overweight (25.0 to 29.9), and obese (≥ 30) [34].

6.2.7   Covariates

Socio-demographic characteristics included gender, age, household size, whether the participant had completed at least primary level education, and whether the participant currently had paid work (yes/no).

6.2.8   Analysis

We conducted descriptive analyses to characterize the sample and then multivariable linear regression analysis examining pre- and post-intervention changes (comparing baseline and 2 month follow-up) in food insecurity, nutritional knowledge, dietary intake (number of times eaten and diversity of diet), and nutritional status (BMI and mid-arm and waist circumference), while controlling for baseline characteristics that could potentially confound the outcomes, including gender, age, education, and work status. Random effects were used to control for the correlations between repeated measures for each participant and the clustering of participants within each clinic. Among 482 participants at baseline, we had follow-up data for 364 (76 %). Of these, 8 were missing data on covariates used in the analyses, thus the final analytic sample was 356 adults. We did not have data on adherence to the nutritional counseling, thus the results represent an intention to treat analysis. We fitted the proposed model using the restricted maximum likelihood, which is a consistent estimator under the missing at random assumption. This approach is particular suitable when most missing data cases are in the outcome variables.

6.3   RESULTS

6.3.1   Baseline Characteristics

Table 2 summarizes the socio-demographic characteristics and primary outcomes for the sample at baseline, first among the final analytic sample (n = 356) and then among those who dropped out or were missing data (n = 126). Most of those in the analytic sample were women (62 %), Mestizo (85 %), and had at least a primary level education (52 %), a median age of 40 years, and a mean household size of 4 people. Labor force participation was generally low, especially among women (22 % reported currently working compared to 53 % of men). Study dropouts were generally similar to the analytic sample, but had a smaller percentage of completely primary level education (45 %), and a BMI distribution more skewed to the left (i.e. towards lower BMI).

TABLE 6.2   Participants' socio-demographic characteristics and baseline measures of study outcomes (n = 356)

Characteristics	Study Completers			Study Dropouts
	Women (n=220)	Men (n=136)	All (n=356)	All (n=126)
Female, %	..	..	61.8	61.2
Ethnic group, %				
Indigenous	2.8	0.8	2.0	1.2
Mestizo	81.6	90.2	84.9	90.5
Afro descendants	14.8	9.0	12.6	8.3
Other/Don't know/Not stated	0.8	0.0	0.5	0.0
Median age [IQR]	39.0 [16.0]	38.0 [15.0]	40.0 [15.0]	39.0 [15.0]
Median household size, n [IQR]	4.0 [2.5]	4.0 [3.0]	4.0 [2.5]	4.0 [2.7]
Actually has work, %	22.3	52.9	34.0	35.7
Completed primary level education, %	46.8	61.0	52.2	44.6
Baseline measures of study outcomes				
Median 8-item food insecurity score[a] [IQR]	6.0 [5.0]	6.0 [5.0]	6.0 [5.0]	6.0 [5.0]
Median 15-item food insecurity score[a] [IQR]	11.0 [9.0]	8.0 [10.0]	9.5 [8.0]	10.0 [8.0]
Median nutritional knowledge score [IQR]	11.0 [2.0]	11.0 [2.0]	11.0 [2.0]	11.0 [4.0]
Median # times ate in past 24 hours [IQR]	3.0 [1.0]	4.0 [1.0]	4.0 [1.0]	3.5 [1.0]
Median # food types at meals (past 24 hrs) [IQR]	2.0 [2.0]	2.0 [1.0]	2.0 [2.0]	2.0 [2.0]

TABLE 6.2 *(Continued)*

Characteristics	Study Completers			Study Dropouts
	Women (n=220)	Men (n=136)	All (n=356)	All (n=126)
Median Body Mass Index [IQR]	23.4 [7.1]	21.6 [5.4]	22.5 [6.6]	21.3 [6.1]
Underweight (<18.5), %	7.3	10.3	8.4	21.4
Normal (18.5-24.9), %	49.1	61.8	54.0	51.8
Overweight (25.0-29.9), %	26.4	19.9	23.9	17.0
Obese (30.0+), %	17.3	8.1	13.8	9.8
Median mid-arm [IQR]	27.0 [6.1]	27.0 [4.0]	27.0 [5.6]	26.0 [4.1]
Median Waist circumference [IQR]	85.0 [17.0]	80.2 [14.0]	83.0 [16.0]	83.0 [16.0]

[a]All participants were asked the 8-item household food insecurity scale. Only those who had children under 18 living in the household were asked the additional 7 items on children and thus are included under the 15-item food insecurity measure.

In terms of outcomes, levels of food insecurity were generally high, with a median of 6 on the 8-item scale and 9 on the 15-item scale, both indicating "moderate" food insecurity. Nutritional knowledge was also generally high, with a median of 11 correctly answered items out of 13. Median number of times eaten in past 24 h was 4 and median number of food types eaten per meal in past 24 h was 2. Median BMI was 23 kg/m^2, while median arm and waist circumferences were 27.4 cm and 83.5 cm, respectively.

6.3.2 Multivariable Regression Analyses (Table 3)

Household food insecurity decreased on average between baseline and follow-up among all participants (n = 356, 8-item scale, $\beta = -0.47$, p < .05) and among those with children under 18 (n = 303, 15-item scale, $\beta = -1.16$, p < .01). Not working currently was associated with greater food insecurity on both scales ($\beta = 0.48$, p < .05 and $\beta = 1.26$, p < .01, respectively), as was not having completed a primary school education ($\beta = 0.95$, p < .001 and $\beta = 2.14$, p < .001, respectively).

Nutritional knowledge also improved between baseline and follow-up ($\beta = 0.88$, p < .001). Not having completed a primary school education was associated with lower nutritional knowledge ($\beta = -0.25$, p < .05).

Dietary intake improved in terms of the number of times participants ate in the past 24 h ($\beta = 0.30$, p < .001) and the average number of different types

TABLE 6.3　Pre-and post-intervention changes in food insecurity and nutritional outcomes[‡]

Variable	Food insecurity (8 items)	Food insecurity (15 items)	Nutritional knowledge (13 items)	# eating occasions in past 24h	# food types at each meal (24h)	BMI	Mid arm circum.	Waist circum.
	N=356	N=303	N=356	N=364	N=364	N=356	N=356	N=356
Nutritional counseling treatment	−0.47 (0.19)*	−1.16 (0.40)**	0.88 (0.11)***	0.30 (0.07)***	0.15 (0.04)***	−0.30 (0.38)	−0.10 (0.40)	−1.46 (0.84)
Female gender	0.20 (0.32)	0.49 (0.71)	0.25 (0.17)	0.03 (0.08)	−0.16 (0.06)*	1.42 (0.42)***	−0.69 (0.63)	1.60 (0.94)
Age	−0.01 (0.01)	0.00 (0.02)	0.00 (0.00)	0.00 (0.00)	0.00 (0.00)	0.03 (0.02)	0.04 (0.02)*	0.12 (0.04)**
Household size	−0.03 (0.04)	0.04 (0.08)	0.00 (0.02)	0.01 (0.01)	0.01 (0.01)	0.14 (0.08)	0.13 (0.08)	0.48 (0.17)**
Not working	0.48 (0.23)*	1.26 (0.47)**	−0.22 (0.13)	−0.16 (0.08)	−0.04 (0.05)	0.42 (0.45)	0.56 (0.47)	0.84 (0.99)
Less than completed primary ed.	0.95 (0.20)***	2.14 (0.41)***	−0.25 (0.11)*	−0.18 (0.07)*	−0.12 (0.04)**	−0.59 (0.40)	−0.26 (0.42)	−1.15 (0.88)

*** p<.001;
** p<.01;
* p<.05
[‡]Nutritional counseling treatment and covariate effects were estimated by mixed-effect linear regression models adjusted for within-participant correlations and fitted by restricted maximum likelihood estimators.

of foods consumed at each meal ($\beta = 0.15$, p < .001). Less than complete primary education was associated with worse dietary intake on these two measures ($\beta = -0.18$, p < .05 and $\beta = -0.12$, p < .01, respectively).

Nutritional status as measured by BMI and mid-arm and waist circumferences showed no significant changes between baseline and 2 month follow-up. Female gender was associated with higher BMI ($\beta = 1.42$, p < .001) and age was associated with higher mid-arm and waist circumferences ($\beta = 0.04$, p < .05 and $\beta = 0.12$, p < .01, respectively). Larger household size was also associated with higher waist circumference ($\beta = 0.48$, p < .01).

6.4   DISCUSSION

A culturally and locally appropriate nutritional counseling intervention delivered by peer counselors was significantly associated with improved nutritional knowledge and dietary intake (diversity and frequency) among PLHIV in Honduras. Furthermore, even after controlling for other indicators of socioeconomic status, the intervention was significantly associated with improved food security. The results fill an important gap in evidence addressing the need for effective, sustainable nutrition education and counseling interventions for PLHIV in resource-limited settings [30]. To our knowledge, ours is the first study to demonstrate that a peer-delivered nutritional counseling intervention for PLHIV in resource-limited settings may improve dietary quality and reduce food insecurity among a population of diverse nutritional statuses (underweight, normal weight, and overweight and obese individuals).

The finding that peer nutrition education was associated with improved food insecurity is especially notable given the persistent negative effects of food insecurity on ART adherence [35] and subsequent immunologic and virologic outcomes [36, 37], as well as its association with higher morbidity [3] and mortality among PLHIV [38, 39]. Our counseling intervention focused not only on how to obtain good nutrition using locally available foods but also how to make the most of existing resources for food, which may have decreased food insecurity. The fact that peers delivered the counseling, sharing from their own experiences, likely made the strategies even more salient. The parent grant, which used randomization to assign individuals to study groups, found that nutritional education (both alone and in concert with a monthly household food basket), when delivered by trained nutritionists, significantly improved food security [40] and adherence across three adherence outcomes (missed clinic appointments, delayed prescription refills, and self-reported missed doses of ART)

[16]. Additional research is needed to determine if peer-delivered nutritional counseling can improve adherence among people on ART.

Peer nutritional education may be especially important in low-resource settings with high prevalence of overweight and obesity among food insecure PLHIV. The parent study found that nutrition education delivered by trained nutritionists significantly improved weight, with overweight and obese patients at baseline *losing* weight and underweight and normal weight patients *gaining* weight over the 12 month follow-up [40]. However, when evaluating the addition of food support to nutritional education, food support had the undesired effect of increasing weight significantly among already overweight and obese patients [40]. Further, food supplementation or aid is usually not sustainable in the long-term and thus may fail to address upstream and downstream health consequences [41]. Thus, a nutritional counseling intervention, in addition to being more scalable and sustainable than food support (particularly when delivered by peer counselors), might be more universally appropriate across settings where wasting or underweight is not the primary concern. Indeed, overweight and obesity among PLHIV is of increasing concern across multiple settings and in particular among those on ART, which can lead to increased risk of other chronic conditions including diabetes and cardiovascular disease [42, 43].

Our study had several limitations. The short follow-up period limited our ability to assess effects of the peer nutritional counseling on nutritional status (BMI, body circumferences) and ART adherence. Further, although dropouts and the analytic sample were similar in observed characteristics, we do not know whether the dropouts were systematically different from study completers in unobserved characteristics (such as morbidity or mortality); if so, this could result in selection bias and affect the estimated effects. Finally, our study is based on a pre-post comparison with only one study arm, thus we are unable to control for any secular trend in outcomes unrelated to the peer nutritional counseling. However, given the relatively short study period, it is unlikely that any such secular trend occurred.

6.5   CONCLUSIONS

Addressing the nutritional needs and food security of PLHIV with feasible, evidence-based, locally-tailored approaches is essential to sustainably improve HIV outcomes. The nutritional education model presented in this paper provides an example of how training peer leaders to deliver HIV-specific nutritional counseling may improve key nutritional outcomes using readily available human capital. This approach has the potential to increase scalability and cultural relevance compared to approaches that rely on professional nutritionists. Our study fills a key gap in the literature on nutrition education and counseling interventions for PLHIV in resource-limited settings and future studies should rigorously test the effectiveness of peer-led nutritional education models on ART adherence and HIV outcomes.

REFERENCES

1. Anderson SA. Core indicators of nutritional state for difficult-to-sample populations. J Nutr. 1990;120(11 Suppl):1555–600.
2. Marcellin F, Boyer S, Protopopescu C, Dia A, Ongolo-Zogo P, Koulla-Shiro S, et al. Determinants of unplanned antiretroviral treatment interruptions among people living with HIV in Yaounde, Cameroon (EVAL survey, ANRS 12-116). Trop Med Int Health. 2008;13(12):1470–8.
3. Weiser SD, Tsai AC, Gupta R, Frongillo EA, Kawuma A, Senkungu J, et al. Food insecurity is associated with morbidity and patterns of healthcare utilization among HIV-infected individuals in a resource-poor setting. AIDS. 2012;26(1):67–75.
4. Franke MF, Murray MB, Munoz M, Hernandez-Diaz S, Sebastian JL, Atwood S, et al. Food insufficiency is a risk factor for suboptimal antiretroviral therapy adherence among HIV-infected adults in urban Peru. AIDS Behav. 2011;15(7):1483–9.
5. Weiser SD, Tuller DM, Frongillo EA, Senkungu J, Mukiibi N, Bangsberg DR. Food insecurity as a barrier to sustained antiretroviral therapy adherence in Uganda. PLoS One. 2010;5(4):e10340.
6. Boyer S, Clerc I, Bonono CR, Marcellin F, Bile PC, Ventelou B. Non-adherence to antiretroviral treatment and unplanned treatment interruption among people living with HIV/AIDS in Cameroon: Individual and healthcare supply-related factors. Soc Sci Med. 2011;72(8):1383–92.
7. Sasaki Y, Kakimoto K, Dube C, Sikazwe I, Moyo C, Syakantu G, et al. Adherence to antiretroviral therapy (ART) during the early months of treatment in rural Zambia: influence of demographic characteristics and social surroundings of patients. Ann Clin Microbiol Antimicrob. 2012;11:34.
8. Weiser SD, Palar K, Frongillo EA, Tsai AC, Kumbakumba E, Depee S, et al. Longitudinal assessment of associations between food insecurity, antiretroviral adherence and HIV treatment outcomes in rural Uganda. AIDS. 2014;28(1):115–20.
9. Gebo KA, Keruly J, Moore RD. Association of social stress, illicit drug use, and health beliefs with nonadherence to antiretroviral therapy. J Gen Intern Med. 2003;18(2):104–11.
10. Kalichman SC, Cherry C, Amaral C, White D, Kalichman MO, Pope H, et al. Health and treatment implications of food insufficiency among people living with HIV/AIDS, Atlanta, Georgia. J Urban Health. 2010;87(4):631–41.

11. Weiser SD, Yuan C, Guzman D, Frongillo EA, Riley ED, Bangsberg DR, et al. Food insecurity and HIV clinical outcomes in a longitudinal study of homeless and marginally housed HIV-infected individuals in San Francisco. AIDS. 2013;27(18):2953–8.

12. Peretti-Watel P, Spire B, Schiltz MA, Bouhnik AD, Heard I, Lert F, et al. Vulnerability, unsafe sex and non-adherence to HAART: evidence from a large sample of French HIV/AIDS outpatients. Soc Sci Med. 2006;62(10):2420–33.

13. WHO. Nutrition Counseling, Care and Support for HIV-Infected Women. Guidelines on HIV-related care, treatment and support for HIV-infected women and their children in resource-constrained settings. Geneva: WHO; 2004.

14. Aberman NL, Rawat R, Drimie S, Claros JM, Kadiyala S. Food security and nutrition interventions in response to the AIDS epidemic: assessing global action and evidence. AIDS Behav. 2014;18:S554–S65.

15. de Pee S, Grede N, Mehra D, Bloem MW. The enabling effect of food assistance in improving adherence and/or treatment completion for antiretroviral therapy and tuberculosis treatment: a literature review. AIDS Behav. 2014;18 Suppl 5:S531–41.

16. Martinez H, Palar K, Linnemayr S, Smith A, Derose KP, Ramirez B, et al. Tailored nutrition education and food assistance improve adherence to HIV antiretroviral therapy: evidence from Honduras. AIDS Behav. 2014;18 Suppl 5:S566–77.

17. Almeida LB, Segurado AC, Duran AC, Jaime PC. Impact of a nutritional counseling program on prevention of HAART-related metabolic and morphologic abnormalities. AIDS Care. 2011;23(6):755–63.

18. Serrano C, Laporte R, Ide M, Nouhou Y, de Truchis P, Rouveix E, et al. Family nutritional support improves survival, immune restoration and adherence in HIV patients receiving ART in developing country. Asia Pac J Clin Nutr. 2010;19(1):68–75.

19. Byron E, Gillespie S, Nangami M. Integrating nutrition security with treatment of people living with HIV: lessons from Kenya. Food Nutr Bull. 2008;29(2):87–97.

20. Ivers L, Chang Y, Jerome J, Freedberg K. Food assistance is associated with improved body mass index, food security and attendance at clinic in an HIV program in central Haiti: a prospective observational cohort study. AIDS Res Ther. 2010;7(33):1–8.

21. Arem H, Nakyanjo N, Kagaayi J, Mulamba J, Nakigozi G, Serwadda D, et al. Peer health workers and AIDS care in Rakai, Uganda: a mixed methods operations research evaluation of a cluster-randomized trial. AIDS Patient Care STDS. 2011;25(12):719–24.

22. Chang LW, Kagaayi J, Nakigozi G, Ssempijja V, Packer AH, Serwadda D, et al. Effect of peer health workers on AIDS care in Rakai, Uganda: a cluster-randomized trial. PLoS One. 2010;5(6):e10923.

23. Gusdal AK, Obua C, Andualem T, Wahlstrom R, Chalker J, Fochsen G, et al. Peer counselors' role in supporting patients' adherence to ART in Ethiopia and Uganda. AIDS Care. 2011;23(6):657–62.

24. Jerome G, Ivers L. Community health workers in health systems strengthening: a qualitative evaluation from rural Haiti. AIDS. 2010;24 Suppl 1:S67–72.

25. Louis C, Ivers LC, Fawzi MCS, Freedberg KA, Castro A. Late presentation for HIV care in central Haiti: factors limiting access to care. AIDS Care. 2007;19(4):487–91.

26. Mukherjee JS, Eustache FE. Community health workers as a cornerstone for integrating HIV and primary healthcare. AIDS Care. 2007;19:S73–82.

27. Fisher JD, Fisher WA. Changing AIDS-risk behavior. Psychol Bull. 1992;111(3):455–74.

28. Fisher JD, Fisher WA, Shupper PA. The information motivation behavioral skills model of HIV preventive behavior. 2nd ed. San Francisco: Jossey Bass; 2009.

29. Eicher-Miller HA, Mason AC, Abbott AR, McCabe GP, Boushey CJ. The effect of food stamp nutrition education on the food insecurity of low-income women participants. J Nutr Educ Behav. 2009;41(3):161–8.

30. Kaye HL, Moreno-Leguizamon CJ. Nutrition education and counselling as strategic interventions to improve health outcomes in adult outpatients with HIV: a literature review. Afr J AIDS Res. 2010;9(3):271–83.

31. WHO, UNAIDS, UNICEF. HIV and Infant Feeding Counseling: A Training Course. Director's Guide. Geneva, Switzerland: WHO; 2000.

32. Melgar-Quiñonez H, Cecilia Alvarez Uribe M, Yanira Fonseca Centeno Z, Bermúdez O, Palma de Fulladolsa P, Fulladolsa A, et al. Psychometric characteristics of the food security scale (ELCSA) applied in Colombia, Guatemala y México. Segurança Alimentar e Nutricional, Campinas. 2010;17(1):48–60.

33. Habicht JP. Standardization of quantitative epidemiological methods in the field. Bol Oficina Sanit Panam. 1974;76(5):375–84. Estandarizacion de metodos epidemiologicos cuantitativos sobre el terreno. spa.

34. BMI Classification. Available from: http://apps.who.int/bmi/index.jsp?introPage=intro_3.html. Accessed Jan. 15, 2015.

35. Singer AW, Weiser SD, McCoy SI. Does food insecurity undermine adherence to antiretroviral therapy? A systematic review. AIDS Behav. 2015;19(8):1510–26.

36. McMahon JH, Wanke CA, Elliott JH, Skinner S, Tang AM. Repeated assessments of food security predict CD4 change in the setting of antiretroviral therapy. J Acquir Immune Defic Syndr. 2011;58(1):60–3.

37. Alexy E, Feldman M, Thomas J, Irvine M. Food insecurity and viral suppression in a cross-sectional study of people living with HIV accessing Ryan White food and nutrition services in New York City. Lancet. 2013;382 Suppl 2:S15.

38. Weiser SD, Fernandes KA, Brandson EK, Lima VD, Anema A, Bangsberg DR, et al. The association between food insecurity and mortality among HIV-infected individuals on HAART. J Acquir Immune Defic Syndr. 2009;52(3):342–9.

39. Anema A, Chan K, Chen Y, Weiser S, Montaner JS, Hogg RS. Relationship between food insecurity and mortality among HIV-positive injection drug users receiving antiretroviral therapy in British Columbia, Canada. PLoS One. 2013;8(5):e61277.

40. Palar K, Kushel M, Frongillo EA, Riley ED, Grede N, Bangsberg D, et al. Food insecurity is longitudinally associated with depressive symptoms among homeless and marginally-housed individuals living with HIV. AIDS Behav. 2015;19(8):1527–34.

41. Yager JE, Kadiyala S, Weiser SD. HIV/AIDS, Food Supplementation and livelihood programs in Uganda: a way forward? PloS One. 2011;6(10):e26117.

42. Keithley JK, Duloy AM, Swanson B, Zeller JM. HIV infection and obesity: a review of the evidence. J Assoc Nurses AIDS Care. 2009;20(4):260–74.

43. Finkelstein JL, Gala P, Rochford R, Glesby MJ, Mehta S. HIV/AIDS and lipodystrophy: implications for clinical management in resource-limited settings. J Int AIDS Soc. 2015;18(1):19033.

Relationship between Food Insecurity and Mortality among HIV-Positive Injection Drug Users Receiving Antiretroviral Therapy in British Columbia, Canada

Aranka Anema, Keith Chan, Yalin Chen, Sheri Weiser, Julio S. G. Montaner, and Robert S. Hogg

7.1 INTRODUCTION

Despite the tremendous benefits of antiretroviral therapy (ART) use on HIV disease progression and survival [1], [2], micro- and macronutrient malnutrition remain strong independent predictors of mortality among HIV-positive individuals in both high and low resource settings [3], [4], [5], [6], [7], [8], [9], [10], [11], [12], [13]. A growing body of evidence suggests that socio-economic determinants may also adversely impact survival among people living with HIV/AIDS [14], [15]. More recently, our study team found that among HIV-positive individuals receiving ART in Canada, being food insecure and underweight was independently associated with a 1.94 increased risk of non-accidental death, compared to being food secure and of normal weight [16]. Poor dietary diversity, a component of food insecurity, has been associated with mortality among ART-naïve individuals in Uganda [17]. These studies suggest that food insecurity warrants prioritization by public health programs and policies for HIV-infected populations.

Illicit drug use is a well-known risk factor for food insecurity and poor nutritional status. Drug addiction alters dietary consumption patterns, leading individuals to eat fewer meals [18], to often skip meals for an entire day [19], and to rely on food distribution services for subsistence [20]. Studies have found that the diets of illicit drug users tend to be calorically insufficient and poor in quality [18], resulting in diverse micro- and macronutrient deficiencies [19], [21], [22], [23], [24]. Food insecurity is theorized to be linked to adverse HIV outcomes through distinct nutritional, mental health and behavioral pathways [25]. Studies in urban HIV-positive populations receiving HIV treatment, including a high proportion of illicit drug users, have found that food insecurity is associated with HIV-related wasting [19], virologic non-suppression [26], [27] and poor immunologic response to ART [27], [28], [29].

To our knowledge, no studies have examined whether food insecurity increases risk of excess death in this population. This study therefore aimed to assess the potential relationship between food insecurity and all-cause mortality among HIV-positive injection drug users (IDU) initiating ART across BC. In light of previous findings regarding the relationship between food insecurity and mortality in the general HIV-population in this setting [16], and theorized nutritional mechanisms linking these [25], this study hypothesized that food insecurity is independently associated with all-cause mortality.

7.2 METHODS

7.2.1 Study Sample: HIV/AIDS Drug Treatment Program (DTP)

Survey data for this analysis were obtained from the provincial HIV/AIDS Drug Treatment Program (DTP) administrative database. In BC, antiretrovirals have been distributed free of charge to HIV-positive individuals since 1986, and coordinated centrally since 1992 by the British Columbia Centre for Excellence in HIV/AIDS (BC-CfE) HIV/AIDS DTP, located at St. Paul's Hospital in Vancouver. Details of the HIV/AIDS DTP have been described elsewhere [1].

In brief, clinical eligibility for receipt of ART in BC is based on guidelines generated by the BC-CfE Therapeutic Guidelines Committee, and have remained consistent with current recommendations by the International AIDS Society – USA since 1996 [30]. Currently, the most frequently prescribed initial triple combination ART regimens in BC consist of a nucleoside reverse transcriptase inhibitor (NRTI) and a nucleotide reverse transcriptase inhibitor or two NRTIs as a backbone, plus either i) a non-nucleoside reverse transcriptase inhibitor (NNRTI), or ii) a protease inhibitor boosted with ritonavir (boosted PI) [31].

Prescribing physicians must complete a drug request form, which acts as a legal prescription, to enroll an eligible individual on ART. HIV/AIDS DTP enrollment forms elicit information about patient socio-demographics, HIV-specific drug history, CD4 cell counts, plasma HIV RNA levels, current drug prescription, and the enrolling physician. Prospective clinical, virologic, immunologic and drug-regimen data are collected on a quarterly basis. Epidemiological studies performed on DTP data form the basis of ongoing revisions to the BC-CfE's province-wide HIV treatment guidelines [31]. Patients provide voluntary written informed consent for the BC-CfE to access electronic medical records for research purposes. Ethical approval for these analyses has been provided by the Providence Health Care/University of British Columbia Research Ethics Board.

7.2.2   Variable Selection

The primary outcome variable of interest was all-cause mortality. This outcome includes individuals who died due to HIV/AIDS-related and non-related causes, including co-morbidities, injuries, accidents, trauma, assaults, drug overdose and suicide. Mortality data was collected on an ongoing basis through physician reports. This data was confirmed through electronic linkage to the British Columbia Division of Vital Statistics registry, which has been shown to capture upwards of 96% of all deaths in the province [32]. Study participants and individuals lost to follow-up were censored on Sept 30, 2011.

7.2.2.1   *Primary Explanatory Variable.*

The primary explanatory variable of interest was food insecurity, captured cross-sectionally at baseline in 1998/1999 as part of the DTP enrollment form for participants newly initiating ART. Food insecurity was measured using an abbreviated version of the Radimer/Cornell scale. The Radimer/Cornell scale assumes that food insecurity comprises four distinct components that are experienced distinctly at household and individual levels. The scale poses questions regarding food depletion and intake (quantitative component), suitability and adequacy of food (qualitative component), feelings of anxiety and deprivation associated with food (psychological component), and patterns of food acquisition and eating (social component) [33], [34], and has been extensively validated in North America [33], [35]. Since the Radimer/Cornell scale was published in 1990, operational definitions and measurements of food insecurity have continued to evolve [36], [37]. The Radimer/Cornell scale understood 'hunger' to be a quantitative component of food insecurity, experienced purely at the individual level [33], [34], and defined as an "inability to acquire or consume an adequate quality or sufficient quantity of food in socially acceptable ways, or the uncertainty that one will be able to do so" [36]. A 2006 review of food insecurity and hunger measures by an expert panel for the United Sates Department of Agriculture (USDA) concluded that hunger should be considered an "indicator and possible consequence of food insecurity," and be measured "distinct from, but in the context of, food insecurity" [37]. Conceptual research in the field of HIV/AIDS further emphasizes the need for multi-dimensional indicators to capture different dimensions (utilization, access, availability, stability) of food insecurity, and for stand-alone indicators to measure sub-components of food insecurity (quantity, quality, safety). Reconsideration of how food insecurity is measured is required to capture the complex experiences of food insecurity among people living with HIV/AIDS, to generate 'actionable' findings for public health programs and policies [38].

The analytic approach to measuring food insecurity and hunger in this study diverged from Radimer/Cornell's and USDA's published methodologies [33], [39]. Participants were categorized as experiencing household food insecurity if they gave a minimum of one positive answer (often/sometimes) to any one of the eight items measuring household food insecurity, which provided a conservative estimate of food insecurity. In the absence of a validated single-question measure of hunger to ascertain food insufficiency at the individual level [37], [38], and in light of public health policy and program requests in our setting to understand what sub-component (quantity, quality, safety) of food insecurity may be driving associations between food insecurity and adverse health outcomes, we extracted and analyzed hunger at the individual level, defined as responding 'often/sometimes' to the question: "I am often hungry, but don't eat because I can't afford enough food" [33], [34].

7.2.2.2   *Secondary Explanatory Variables.*

Secondary explanatory variables hypothesized to confound the relationship between food insecurity and mortality were selected based on findings from previous literature, including a previous study examining the impacts of food insecurity on mortality among HIV-positive individuals in this setting [16]. Socio-demographic variables included: age at ART start date (continuous); gender (male vs. female); Aboriginal ancestry (yes vs. no); annual income (>CAD$15,000 vs. ≤CAD$15,000), with a dichotomous split based on Canada Revenue Agency's low income threshold [40]; education (>high school vs. ≤high school graduation); and unstable housing (yes vs. no), defined as living in a hotel, boarding house, group home, jail, on the street, or having no fixed address at the time of the survey. Clinical variables considered in this analysis included nutritional status, measured by physician and self-reported body mass index (BMI) (kg/m^2) and calculated using the formula: weight/(height)^2. Cut-offs for underweight status were based on current WHO standards for HIV-positive individuals, defined as <18.5 kg/m^2 (underweight) vs. ≥18.5 kg/m^2 (not underweight) [41]. BMI reflects lean body mass and fat mass, has been shown to detect malnutrition at an earlier stage than other anthropometric measures, and is considered a sensitive screening tool for malnutrition among people living with HIV/AIDS [42]. Other clinical variables included use of triple combination highly active antiretroviral therapy (HAART) (yes vs. no); PI-based regimen (yes vs. no); AIDS diagnosis (yes vs. no), defined according to CDC classification [43]; plasma HIV RNA viral load (per log10 copies/mL), measured at most recent date prior to survey; CD4 cell counts (per 100 cells/µL), recorded at most recent date prior to survey using flow cytometry and fluorescent monoclonal antibody analysis (Beckman Coulter, Inc., Mississauga, Ontario, Canada); and finally ART adherence, measured on the basis of prescription refill compliance [44], defined as the number of days ART was dispensed over the number of days an individual was eligible for ART in the past 12 months (≥95% vs. <95%), at most recent date prior to survey. This adherence variable has shown to reliably predict survival among IDU in previous studies [44], [45]. All potential explanatory variables were collected at baseline in 1998/1999 from the HIV/AIDS DTP enrollment survey, except where indicated above.

7.2.3   Statistical Analysis

As a first step, bivariate analyses were performed on the entire study sample at baseline, stratified by food insecurity, hunger and all-cause mortality, respectively. Pearson's Chi-Square tests were used to compare categorical variables. In instances where counts were small (five or less), the Fisher's Exact Test was used. Continuous variables were compared using Wilcoxon Rank Sum Test. Next, two Cox proportional hazard confounder models were constructed to determine the association between food insecurity/hunger and all-cause mortality, controlling for potential confounders. A multivariate model was built using an adaptation of methods described by Greenland and colleagues [46], [47]. This manual backward stepwise approach involved first fitting a full model, including all explanatory variables, and noting the value of the coefficient associated with food insecurity/hunger. Reduced models were then constructed, each removing one secondary explanatory variable from the full set. Comparing the value of the coefficient for food insecurity in the full model and each of the reduced models, secondary variables were removed corresponding to the smallest relative change in the coefficient for food insecurity. This iterative process continued until the maximum change of the value for food insecurity from the full model exceeded 5%. The intent of this model building strategy was to retain secondary variables in the final multivariate model with greater relative influence on the relationship between food insecurity and mortality. This technique has been previously applied in studies of HIV-positive individuals to estimate the independent relationship between a hypothesized predictor variable and clinical outcome [48], [49]. As a sub-analysis, this process was repeated to examine the relationship between hunger and mortality. Ad hoc tests to assess potential cohort effects on survival were not deemed necessary since participants were all recruited after the introduction of ART, which led to a homogenous trend in reductions of HIV-related mortality over time [50]. All statistical analyses were completed using R v2.10.1 (R Foundation, Vienna, Austria).

7.3   RESULTS

Baseline characteristics of IDU enrolled in the BC-wide HIV/AIDS DTP in years 1998/1999, stratified by food security status, are show in Table 1. A total of 254 individuals enrolled at baseline, responded to the food security scale question in the survey, and had consistent follow-up during the years under study. Of this analytic sample, 181 (71.26%) reported being food insecure, and 108

(42.5%) were hungry; the median age was 38.0 years [interquartile range [IQR]: 34.0–43.0]; 211 (83.07%) were male; 58 (22.9%) reported Aboriginal ancestry; and 219 (96.9%) had a BMI above 18.5 kg/m². The median CD4 cell count was 380.0 per 100 cells/μL (IQR: 220.0–510.0); the median viral load was 2.6 log10 copies/mL (IQR: 2.6–3.7); 63 (24.8%) were diagnosed with AIDS; and 123 (48.4%) were ≥95% adherent to ART in the 12 months preceding enrollment. During the study period (between June 21, 1998 and Sept 30, 2011), a total of 105 (41.34%) individuals died. Bivariate comparison of participant characteristics by food security status revealed that individuals with lower incomes, receiving a protease inhibitor (PI)-based regimen, and initiating ART at a later year, were all significantly more likely to be food insecure (p<0.05). Median follow-up time was 140.61 months (IQR: 59.63–151.20), and median survival time was 61.04 (IQR: 31.11–83.02). A total of 87 (48.1%) individuals who reported being food insecure died over the study period, compared to 18 (24.7%) individuals who reported being food secure (p=0.001). A total of 54 (51.4%) individuals reporting hunger died, compared to 54 (36.2%) reporting no hunger (p=0.022). Product limit survival estimates revealed that the death rate of the 254 IDU who answered the food security and hunger questions on the baseline survey did not differ from other IDU initiating ART during the same time period.

TABLE 7.1 Baseline characteristics among HIV-positive injection drug users initiating antiretroviral therapy across British Columbia, by food security status, between June 1998 and Sept 2011 (n=254).

Characteristic	Total N (%)	Food Insecure 181 (71.3%)	Food Secure 73 (28.7%)	p - value
Age				
Median, IQR[1]	38.0 (34.0–43.0)	38.0 (34.0–43.0)	38.0 (34.0–43.0)	0.933
Gender				
Male	211 (83.1%)	148 (81.8%)	63 (86.3%)	0.492
Female	43 (16.9%)	33 18.2%)	10 (13.7%)	
Aboriginal ancestry				
Yes	58 (22.9%)	45 (25.0%)	13 (17.8%)	0.286
No	195 (77.1%)	135 (75.0%)	60 (82.2%)	
Unstable housing				
Yes	28 (11.8%)	21 (12.2%)	7 (10.8%)	0.936
No	209 (88.2%)	151 (87.8%)	58 (89.2%)	
Education status				
≥High school	161 (64.7%)	110 (61.5%)	51 (72.9%)	0.122
<High school	88 (35.3%)	69 (38.5%)	19 (27.1%)	

TABLE 7.1 *(Continued)*

Characteristic	Total N (%)	Food Insecure 181 (71.3%)	Food Secure 73 (28.7%)	p - value
Annual income				
>$15,000	63 (28.1%)	23 (14.7%)	40 (58.8%)	,0.001
≤$15,000	161 (71.9%)	133 (85.3%)	28 (41.2%)	
Body Mass Index				
≥18.5 kg/m2	219 (96.9%)	153 (96.2%)	66 (98.5%)	0.629
<18.5 kg/m2	7 (3.1%)	6 (3.8%)	1 (1.5%)	
AIDS diagnosis				
Yes	63 (24.8%)	44 (24.3%)	19 (26.0%)	0.899
No	191 (75.2%)	137 (75.7%)	54 (74.0%)	
ART start year				
Median, IQR[1]	1997 (1995–1998)	1997 (1996–1998)	1996 (1994–1998)	0.015
HAART use[2]				
Yes	191 (75.2%)	138 (76.2%)	53 (72.6%)	0.655
No	63 (24.8%)	43 (23.8%)	20 (27.4%)	
PI-based regimen[3]				
Yes	154 (60.6%)	101 (55.8%)	53 (72.6%)	0.019
No	100 (39.4%)	80 (44.2%)	20 (27.4%)	
Adherence to ART[4]				
≥95%	123 (48.4%)	82 (45.3%)	41 (56.2%)	0.153
<95%	131 (51.6%)	99 (54.7%)	32 (43.8%)	
CD4 cell count (per 100 cells/mL)				
Median, IQR[1]	380 (220–510)	360 (210–500)	400 (230–555)	0.149
Plasma HIV RNA (per log10 copies/mL)				
Median, IQR[1]	2.6 (2.6–3.7)	2.6 (2.6–3.8)	2.6 (2.6–3.2)	0.250
All-Cause Mortality				
Yes	105 (41.3%)	87 (48.1%)	18 (24.7%)	0.001
No	149 (58.7%)	94 (51.9%)	55 (75.3%)	

[1]Inter-quartile range
[2]Highly active antiretroviral therapy use
[3]Protease Inhibitor-based regimen
[4]Within last 12 months of interview
doi:10.1371/journal.pone.0061277.t001

Unadjusted and adjusted analyses of factors associated with mortality among IDU are presented in Table 2. In unadjusted analyses (Column 1), participants who were food insecure were more than twice as likely to die, compared to individuals who were food secure (hazard ratio [HR]=2.41, 95% Confidence Interval [CI]: 1.45–4.01). Participants who were hungry were almost twice as likely to die (HR=1.78, 95% CI: 1.21–2.61) compared to individuals with no hunger. Other factors significantly associated with all-cause mortality included Aboriginal ancestry, low income, <95% adherence to ART, lower median CD4 cell count and higher median plasma HIV RNA. In adjusted analyses, controlling for potential confounders, food insecurity remained significantly associated with all-cause mortality (adjusted hazard ratio [AHR]=1.95, 95% CI: 1.07–3.53) (Column 2), and hunger was no longer significant (AHR=1.05, 95% CI: 0.65–1.70) (Column 3). Additional factors associated with increased likelihood of death in the model examining food insecurity included having an annual income >$15,000, Aboriginal ancestry, and a higher median viral load. In the model exploring hunger, additional characteristics associated with higher risk of mortality included older age, annual income >$15,000 and a higher median viral load.

TABLE 7.2   Unadjusted and adjusted factors associated with all-cause mortality among HIV-positive injection drug users initiating highly active antiretroviral therapy in British Columbia, between June 1998 and Sept 2011 (n=254).

Characteristic	Unadjusted model HZ[1] (95% CI)[2]	Adjusted model including food insecurity AHZ[3] (95% CI)[2]	Adjusted model including hunger AHZ[3] (95% CI)[2]
Food insecure			
Yes vs. no	2.41 (1.45–4.01)	1.95 (1.07–3.53)	–
Hunger			
Yes vs. no	1.78 (1.21–2.61)	–	1.05 (0.65–1.70)
Age			
Per 10 year increase	1.19 (0.94–1.50)	1.27 (0.98–1.65)	1.46 (1.09–1.94)
Gender			
Male vs. female	0.64 (0.41–1.01)	–	0.59 (0.34–1.02)
Aboriginal ancestry			
Yes vs. no	1.92 (1.27–2.92)	2.15 (1.34–3.45)	–
Unstable housing			
Yes vs. no	1.50 (0.87–2.60)	–	0.92 (0.46–1.82)

TABLE 7.2 (*Continued*)

Characteristic	Unadjusted model HZ[1] (95% CI)[2]	Adjusted model including food insecurity AHZ[3] (95% CI)[2]	Adjusted model including hunger AHZ[3] (95% CI)[2]
Education status			
>High school vs. ≤high school	0.85 (0.57–1.25)	–	–
Annual income			
<$15,000 vs. ≤$15,000	0.27 (0.14–0.50)	0.33 (0.16–0.68)	0.28 (0.14–0.58)
AIDS diagnosis			
Yes vs. no	0.95 (0.61–1.49)	–	–
Body Mass Index			
≥18.5 kg/m² vs. <18.5 kg/m²	0.74 (0.27–2.01)	–	–
ART start year			
Per year increase	1.03 (0.93–1.14)	0.88 (0.78–1.00)	0.93 (0.81–1.06)
HAART use[5]			
Yes vs. no	1.21 (0.76–1.93)	–	–
PI-based regimen[6]			
Yes vs. no	0.69 (0.47–1.01)	–	–
Adherence to ART[7]			
≥95% vs. <95%	0.59 (0.40–0.87)	–	0.77 (0.48–1.22)
CD4 cell count			
Per 100 increase	0.89 (0.82–0.98)	0.96 (0.87–1.06)	0.96 (0.86–1.07)
HIV RNA viral load			
Per Log10 increase	1.50 (1.23–1.84)	1.42 (1.12–1.80)	1.36 (1.05–1.75)

[1]Hazard Ratio
[2]95% Confidence Interval
[3]Adjusted Hazard Ratio
[4]Highly active antiretroviral therapy
[5]Protease Inhibitor
[6]Within the last 12 months of interview doi:10.1371/journal.pone.0061277.t002

Adjusted Kaplan-Meir survival probabilities for IDU in this BC-wide cohort, stratified by food insecurity and hunger status, are presented in Figures 1 and 2, respectively. Consistent with findings from the multivariate analyses, individuals reporting food insecurity had significantly reduced probability of survival, compared to individuals who were food secure; and individual reporting hunger did not have lower probability of survival compared to individuals reporting no hunger.

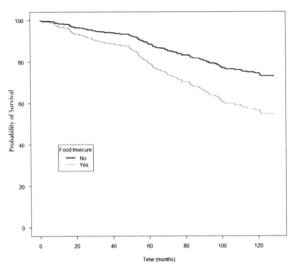

FIGURE 7.1   Adjusted cumulative incidence of all-cause mortality among HIV-positive injection drug users initiating antiretroviral therapy in British Columbia, stratified by food security status.

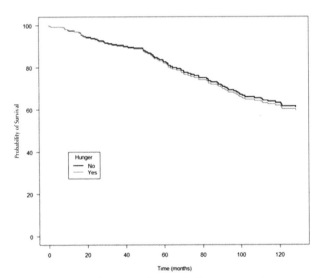

FIGURE 7.2   Adjusted cumulative incidence of all-cause mortality among HIV-positive injection drug users initiating antiretroviral therapy in British Columbia, stratified by hunger status.

7.4   DISCUSSION

This study builds on our existing body of research regarding the relationship between food insecurity and mortality among people living with HIV/AIDS [16]. This study is the first to examine the potential impact of food insecurity and hunger on mortality among HIV-positive IDU. Mortality rates were elevated in this sample of HIV-positive IDU. After 13.3 years of follow-up, individuals who reported being food insecure at baseline were almost twice as likely to die, when controlling for potential confounders. Hunger was associated with increased risk of death in univariate analysis, but the association was no longer significant after controlling for potential confounders in the adjusted analyses. Our results suggest that addressing food insecurity, in addition to other known social and structural barriers to ART adherence and virologic suppression among illicit drug users [51], [52], such as incarceration [53], homelessness [48], and gender-related factors [54], may be of paramount public health importance.

The finding that food insecurity, and not hunger, was significantly associated with all-cause mortality suggests that other aspects of food insecurity, measured within the Radimer/Cornell scale, may be driving this association, including poor dietary diversity and/or anxiety regarding food access. Food insecurity is theorized to be linked to adverse HIV-related outcomes through diverse nutritional, mental health and behavioral pathways [25]. Macro- and micronutrient deficiencies, which represent an extreme consequence of, and surrogate markers for, poor dietary diversity, have been associated with mortality among HIV-positive individuals in the ART era. Compared to non-IDU, active IDU may be at increased risk of developing nutrient deficiencies due to metabolic abnormalities and nutrient interactions associated with drug use[55]. Low albumin and phosphate levels have been associated with early mortality among malnourished individuals initiating ART [7], [9]. Deficiencies in zinc, Vitamin A, iron and B12have also been associated with increased risk of HIV-related mortality in illicit drug-using populations [5], [13], [55], [56]. In particular, selenium deficiency is associated with a 20-fold increase in risk of mortality among HIV-positive IDU [55]. Findings from the current study, taken together with existing evidence regarding the relationship between nutrient deficiencies and mortality among HIV-positive IDU populations, suggest an urgent need for public health bodies to evaluate the possible role of screening for malnutrition and food supplementation to prevent excess mortality among IDU receiving ART in this setting.

The observed association between food insecurity and all-cause mortality in this sample of IDU receiving ART may also be explained by mental health mechanisms. Two questions on the Radimer/Cornell scale pertain specifically to feelings of anxiety regarding food access: "I worry whether my food will run out before I get money to buy more" and "I worry about whether the food that I can afford to buy for my household will be enough." Obsessive or chronic worrying is recognized as a symptom of generalized anxiety disorder within the DSM-V [57]. Mental health disorders, including anxiety and depression, are commonly reported among HIV-positive populations, and believed to increase risk of mortality through both behavioral and biologic mechanisms. Symptoms of depression have been associated with poor virologic response [58], [59], [60], reduced immunologic capacity [61], and AIDS and non-AIDS related death among individuals on ART [49], [62], [63], [64], [65]. Feelings of guilt, fear and discrimination have been associated with delayed access to HIV treatment and care [66], and non-adherence to ART [67], [68], [69]. A recent study of 9,003 HIV-positive individuals in the US found that presence of mental health disorders, including schizophrenia and bi-polar disorder, were significantly associated with all-cause mortality [70]. Taken together with this previous evidence, findings from this study suggest a need to evaluate the possible role of comprehensive mental health support in the context of existing harm reduction and HIV treatment services in order to prevent excess mortality.

This study has several strengths and limitations that warrant consideration. Participants were not randomly selected, and therefore are not representative of the general HIV-positive or IDU populations in BC. Common to all survival analyses, the censoring of participants who were either event-free or lost to follow-up may have led to an underestimation of true time to event. Longitudinal data was not available for most socio-demographic and clinical variables. A major limitation of the study is that food security and hunger status were only collected at baseline. This study was therefore unable to ascertain the possible time-updated effects of food insecurity/hunger on mortality. Multivariate models in this study controlled for several variables that have been hypothesized to be on the causal pathway between food insecurity and mortality [25], and may have therefore underestimated true effect size. Information bias, and specifically responder bias, may have led to non-differential misclassification of hunger status, biasing Odds Ratio estimates towards the null. Residual confounding – due to dichotomization of continuous variables, use of surrogate markers, misclassification, or failure to account for unobserved/unknown confounders – may have introduced bias into effect estimates. Future studies could be strengthened by considering the potential confounding impact of geographic region on the relationship between hunger and mortality, which has been independently associated with both HIV-related food insecurity (data not published) and mortality trends in BC[50].

Although the Radimer/Cornell scale is not considered the most contemporary of food insecurity measurement options, select measures within the scale have been incorporated into contemporary food security scales, including the United States Household Food Security Survey Module (HFSSM) [71] and Canadian versions of the module [72], offering some degree of geographic and temporal comparability. Notably, the hunger measure used in our analysis has remained consistent across different food security modules developed since the early 1990s, with the only difference being an emphasis on hunger frequency and duration in contemporary tools [37]. The advantage of the Radimer/Cornell scale used in this study is that it prompted respondents for 'current' food insecurity and hunger status, minimizing recall bias.

Because survey data were self-reported, this study may have also been susceptible to recall bias and social desirability bias. While the measure of hunger used in this study has been extensively validated in low income populations [27], [73], [74], future studies could be strengthened by applying robust dietary intake assessment methods validated for use among HIV-positive individuals, including 24 hour dietary recall and food frequency questionnaires [75],[76]. Use of self-reported weight and height for BMI may have been prone to bias, with studies demonstrating that individuals tend to under-report energy intake, weight and BMI, and over-report height [77]. Future studies could additionally consider clinical assessment of body weight loss and body cell mass, and subjective assessments of global nutritional status, which have found to be robust among people living with HIV/AIDS [42], and body composition and biochemical measures commonly used in HIV-positive IDU populations [76]. In summary, we found that food insecurity was associated with a two-fold increased risk of mortality, after controlling for potential confounders, among HIV-positive IDU receiving ART across BC, Canada. Further research is necessary to understand the mechanisms through which food insecurity drives this association, and to examine modifying effects of diverse nutritional, mental health and behavioral factors in the relationship between food insecurity and mortality. Public health organizations should prospectively evaluate the possible role of food supplementation and socio-structural supports on survival among IDU within HIV treatment programs.

REFERENCES

1. Hogg RS, Yip B, Chan KJ, Wood E, Craib KJ, et al. (2001) Rates of disease progression by baseline CD4 cell count and viral load after initiating triple-drug therapy. JAMA 286: 2568–2577. doi: 10.1001/jama.286.20.2568

2. Hogg RS, Heath KV, Yip B, Craib KJ, O'Shaughnessy MV, et al. (1998) Improved survival among HIV-infected individuals following initiation of antiretroviral therapy. JAMA 279: 450–454. doi: 10.1001/jama.279.6.450

3. van der Sande MA, Schim van der Loeff MF, Aveika AA, Sabally S, Togun T, et al. (2004) Body mass index at time of HIV diagnosis: a strong and independent predictor of survival. J Acquir Immune Defic Syndr 37: 1288–1294. doi: 10.1097/01.qai.0000122708.59121.03

4. Tang AM, Forrester J, Spiegelman D, Knox TA, Tchetgen E, et al. (2002) Weight loss and survival in HIV-positive patients in the era of highly active antiretroviral therapy. J Acquir Immune Defic Syndr 31: 230–236. doi: 10.1097/00126334-200210010-00014

5. Argemi X, Dara S, You S, Mattei JF, Courpotin C, et al. (2012) Impact of malnutrition and social determinants on survival of HIV infected adults starting antiretroviral therapy in a rural HIV care centre in Cambodia. AIDS 26: 1161–1166. doi: 10.1097/qad.0b013e328353f363

6. Liu E, Spiegelman D, Semu H, Hawkins C, Chalamilla G, et al. (2011) Nutritional status and mortality among HIV-infected patients receiving antiretroviral therapy in Tanzania. J Infect Dis 204: 282–290. doi: 10.1093/infdis/jir246

7. Koethe JR, Blevins M, Nyirenda C, Kabagambe EK, Shepherd BE, et al. (2011) Nutrition and inflammation serum biomarkers are associated with 12-week mortality among malnourished adults initiating antiretroviral therapy in Zambia. J Int AIDS Soc 14: 19. doi: 10.1186/1758-2652-14-19

8. Koethe JR, Limbada MI, Giganti MJ, Nyirenda CK, Mulenga L, et al. (2010) Early immunologic response and subsequent survival among malnourished adults receiving antiretroviral therapy in Urban Zambia. AIDS 24: 2117–2121. doi: 10.1097/qad.0b013e32833b784a

9. Heimburger DC, Koethe JR, Nyirenda C, Bosire C, Chiasera JM, et al. (2010) Serum phosphate predicts early mortality in adults starting antiretroviral therapy in Lusaka, Zambia: a prospective cohort study. PloS One 5: e10687. doi: 10.1371/journal.pone.0010687

10. Moh R, Danel C, Messou E, Ouassa T, Gabillard D, et al. (2007) Incidence and determinants of mortality and morbidity following early antiretroviral therapy initiation in HIV-infected adults in West Africa. AIDS 21: 2483–2491. doi: 10.1097/qad.0b013e3282f09876

11. Paton NI, Sangeetha S, Earnest A, Bellamy R (2006) The impact of malnutrition on survival and the CD4 count response in HIV-infected patients starting antiretroviral therapy. HIV Med 7: 323–330. doi: 10.1111/j.1468-1293.2006.00383.x

12. Zachariah R, Fitzgerald M, Massaquoi M, Pasulani O, Arnould L, et al. (2006) Risk factors for high early mortality in patients on antiretroviral treatment in a rural district of Malawi. AIDS 20: 2355–2360. doi: 10.1097/qad.0b013e32801086b0

13. Semba RD, Graham NM, Caiaffa WT, Margolick JB, Clement L, et al. (1993) Increased mortality associated with Vitamin A deficiency during human immunodeficiency virus type 1 infection. Arch Intern Med 153: 2149–2154. doi: 10.1001/archinte.1993.00410180103012

14. Cunningham WE, Hays RD, Duan N, Andersen R, Nakazono TT, et al. (2005) The effect of socioeconomic status on the survival of people receiving care for HIV infection in the United States. J Health Care Poor Underserved 16: 655–676. doi: 10.1353/hpu.2005.0093

15. McMahon J, Wanke C, Terrin N, Skinner S, Knox T (2011) Poverty, hunger, education, and residential status impact survival in HIV. AIDS Behav 15: 1503–1511. doi: 10.1007/s10461-010-9759-z

16. Weiser SD, Fernandes KA, Brandson EK, Lima VD, Anema A, et al. (2009) The association between food insecurity and mortality among HIV-infected individuals on HAART. J Acquir Immune Defic Syndr 52: 342–349. doi: 10.1097/qai.0b013e3181b627c2

17. Rawat R, McCoy SI, Kadiyala S (2012) Poor diet quality is associated with low CD4 count and anemia and predicts mortality among antiretroviral therapy naive HIV-positive adults in Uganda. J Acquir Immune Defic Syndr 62: 246–53. doi: 10.1097/qai.0b013e3182797363

18. Himmelgreen DA, Pérez-Escamilla R, Segura-Millán S, Romero-Daza N, Tanasescu M, et al. (1998) A comparison of the nutritional status and food security of drug-using and non-drug-using Hispanic women in Hartford, Connecticut. Am J Phys Anthropol 107: 351–361. doi: 10.1002/(sici)1096-8644(199811)107:3<351::aid-ajpa10>3.0.co;2-7

19. Campa A, Yang Z, Lai S, Xue L, Phillips JC, et al. (2005) HIV-Related Wasting in HIV-Infected Drug Users in the Era of Highly Active Antiretroviral Therapy. Clin Infect Dis 41: 1179–1185. doi: 10.1086/444499

20. Romero-Daza N, Himmelgreen DA, Pérez-Escamilla R, Segura-Millán S, Singer M (1999) Food habits of drug-using Puerto Rican women in inner-city Hartford. Medical Anthropology: Cross-Cultural Studies in Health and Illness 18: 281–298. doi: 10.1080/01459740.1999.9966158

21. Santolaria-Fernández FJ, Gómez-Sirvent JL, González-Reimers CE, Batista-López JN, Jorge-Hernández JA, et al. (1995) Nutritional assessment of drug addicts. Drug Alcohol Depend 38: 11–18. doi: 10.1016/0261-5614(93)90055-9

22. Quach LA, Wanke CA, Schmid CH, Gorbach SL, Mwamburi DM, et al. (2008) Drug use and other risk factors related to lower body mass index among HIV-infected individuals. Drug Alcohol Depend 95: 30–36. doi: 10.1016/j.drugalcdep.2007.12.004

23. Nazrul Islam SK, Jahangir Hossain K, Ahmed A, Ahsan M (2002) Nutritional status of drug addicts undergoing detoxification: prevalence of malnutrition and influence of illicit drugs and lifestyle. British Journal of Nutrition 88: 507–513. doi: 10.1079/bjn2002702

24. Forrester JE, Tucker KL, Gorbach SL (2005) The effect of drug abuse on body mass index in Hispanics with and without HIV infection. Public Health Nutr 8: 61–68. doi: 10.1079/phn2005667

25. Weiser SD, Young SL, Cohen CR, Kushel MB, Tsai AC, et al. (2011) Conceptual framework for understanding the bidirectional links between food insecurity and HIV/AIDS. Am J Clin Nutr 94: 1729S–1739S. doi: 10.3945/ajcn.111.012070

26. Weiser SD, Frongillo EA, Ragland K, Hogg RS, Riley ED, et al. (2008) Food Insecurity is Associated with Incomplete HIV RNA Suppression Among Homeless and Marginally Housed HIV-infected Individuals in San Francisco. J Gen Intern Med 24: 14–20. doi: 10.1007/s11606-008-0824-5

27. Kalichman SC, Cherry C, Amaral C, White D, Kalichman MO, et al. (2010) Health and treatment implications of food insufficiency among people living with HIV/AIDS, Atlanta, Georgia. J Urban Health 87: 631–641. doi: 10.1007/s11524-010-9446-4

28. Normén L, Chan K, Braitstein P, Anema A, Bondy G, et al. (2005) Food insecurity and hunger are prevalent among HIV-positive individuals in British Columbia, Canada. J Nutr 135: 820–825.

29. Weiser SD, Bangsberg DR, Kegeles SK, Ragland K, Kushel MB, et al. (2009) Food insecurity among homeless and marginally housed individuals living with HIV/AIDS in San
Francisco. AIDS Behav 13: 841–848. doi: 10.1007/s10461-009-9597-z

30. Thompson MA, Aberg JA, Cahn P, Montaner JS, Rizzardini G, et al. (2010) Antiretroviral
treatment of adult HIV infection: 2010 recommendations of the International AIDS
Society-USA panel. JAMA 304: 321–333. doi: 10.1001/jama.2010.1004

31. British Columbia Centre for Excellence in HIVAIDS (BC-CfE) (Jan. 2011) Therapeutic
Guidelines: Antiretroviral treatment (ARV) of adult HIV infection. Available: http://www.
cfenet.ubc.ca/our-work/initiatives/therapeutic-guidelines/adult-therapeutic-guidelines.
Accessed 2012 Feb 7.

32. Au-Yeung CG, Anema A, Chan K, Yip B, Montaner JS, et al. (2010) Physician's manual
reporting underestimates mortality: evidence from a population-based HIV/AIDS treatment program. BMC Public Health 10: 642. doi: 10.1186/1471-2458-10-642

33. Kendall A, Olson CM, Frongillo EA (1995) Validation of the Radimer/Cornell Measures of
Hunger and Food Insecurity. J Nutr 125: 2793–2801.

34. Radimer KL, Olson CM, Campbell CC (1990) Development of indicators to assess hunger.
J Nutr 120 Suppl 11: 1544–1548.

35. Frongillo EA Jr (1999) Validation of measures of food insecurity and hunger. J Nutr 129:
506S–509S.

36. Radimer KL, Radimer KL (2002) Measurement of household food security in the USA and
other industrialised countries. Public Health Nutr 5: 859–864. doi: 10.1079/phn2002385

37. United States Department of Agriculture (USDA) (2006) Food insecurity and hunger in
the United States: An assessment of the measure. Panel to Review the U.S. Department of
Agriculture's Measurement of Food Insecurity and Hunger. Wunderlich GS and Norwood
JL Eds. The National Academies Press. Washington DC. Available: http://www.nap.edu/
catalog/11578.html. Accessed 2013 Feb 19.

38. Anema A (2011) Food insecurity in the context of HIV: Towards harmonized definitions
and indicators International Conference on AIDS and STIs in Africa Addis Ababa Dec
4–8.

39. Bickel D, Nord M, Price C, Hamilton W, Cook J (2000) United States Department of
Agriculture (USDA) Guide to Measuring Household Food Security. Revised 2000.
Available: www.fns.usda.gov/fsec/FILES/FSGuide.pdf. Accessed 2013 Feb 19.

40. Statistics Canada, Income Statistics Division (2011) Low Income Lines, 2009–2010.
Income Research Paper Series. Catalogue no. 75F0002M — No. 002.

41. World Health Organisation (WHO) (2012) BMI Classification. Table 1: The
International Classification of adult underweight, overweight and obesity according to
BMI. Global Databse on Body Mass Index. Available: http://apps.who.int/bmi/index.
jsp?introPage=intro_3.html. Accessed 2012 May 5.

42. Niyongabo T, Melchior JC, Henzel D, Bouchaud O, Larouzé B (1999) Comparison of
methods for assessing nutritional status in HIV-infected adults. Nutrition 15: 740–743. doi:
10.1016/s0899-9007(99)00146-x

43. Centres for Disease Control (CDC) (1992) 1993 revised classification system for HIV
infection and expanded surveillance case definition for AIDS among adolescents and adults.
Morbidity and Mortality Weekly Report (MMWR) 41: 1–19. doi: 10.1001/jama.269.6.729

44. Wood E, Hogg RS, Yip B, Harrigan PR, O'Shaughnessy MV, et al. (2003) Effect of medication adherence on survival of HIV-infected adults who start highly active antiretroviral

therapy when the CD4+ cell count is 0.200 to 0.350×10(9) cells/L. Ann Intern Med 139: 810–816. doi: 10.7326/0003-4819-139-10-200311180-00008

45. Wood E, Hogg RS, Lima VD, Kerr T, Yip B, et al. (2008) Highly active antiretroviral therapy and survival in HIV-infected injection drug users. JAMA 300: 550–554. doi: 10.1001/jama.300.5.550

46. Maldonado G, Greenland S (1993) Simulation study of confounder-selection strategies. Am J Epidemiol 138: 923–936.

47. Rothman KJ, Greenland S (1998) Modern Epidemiology. New York, NYUnited States: Lippincott Williams & Wilkins.

48. Milloy MJ, Kerr T, Bangsberg DR, Buxton J, Parashar S, et al. (2012) Homelessness as a structural barrier to effective antiretroviral therapy among HIV-seropositive illicit drug users in a Canadian setting. AIDS Patient Care STDS 26: 60–67. doi: 10.1089/apc.2011.0169

49. Lima VD, Geller J, Bangsberg DR, Patterson TL, Daniel M, et al. (2007) The effect of adherence on the association between depressive symptoms and mortality among HIV-infected individuals first initiating HAART. AIDS 21: 1175–1183. doi: 10.1097/qad.0b013e32811ebf57

50. Lima VD, Lepik KJ, Zhang W, Muldoon KA, Hogg RS, et al. (2010) Regional and temporal changes in HIV-related mortality in British Columbia, 1987–2006. Can J Public Health 101: 415–419.

51. Nolan S, Milloy MJ, Zhang R, Kerr T, Hogg RS, et al. (2011) Adherence and plasma HIV RNA response to antiretroviral therapy among HIV-seropositive injection drug users in a Canadian setting. AIDS Care 23: 980–987. doi: 10.1080/09540121.2010.543882

52. Wood E, Montaner JS, Yip B, Tyndall MW, Schechter MT, et al. (2003) Adherence and plasma HIV RNA responses to highly active antiretroviral therapy among HIV-1 infected injection drug users. CMAJ 169: 656–661.

53. Milloy MJ, Kerr T, Buxton J, Rhodes T, Guillemi S, et al. (2011) Dose-response effect of incarceration events on nonadherence to HIV antiretroviral therapy among injection drug users. J Infect Dis 203: 1215–1221. doi: 10.1093/infdis/jir032

54. Tapp C, Milloy MJ, Kerr T, Zhang R, Guillemi S, et al. (2011) Female gender predicts lower access and adherence to antiretroviral therapy in a setting of free healthcare. BMC Infect Dis 11: 86. doi: 10.1186/1471-2334-11-86

55. Baum MK (2000) Role of micronutrients in HIV-infected intravenous drug users. J Acquir Immune Defic Syndr 25: S49–52. doi: 10.1097/00042560-200010001-00008

56. Tang AM, Graham NM, Chandra RK, Saah AJ (1997) Low serum Vitamin B-12 concentrations are associated with faster human immunodeficiency virus type 1 (HIV-1) disease progression. J Nutr 127: 345–351.

57. Andrews G, Hobbs MJ, Borkovec TD, Beesdo K, Craske MG, et al. (2010) Generalized worry disorder: a review of DSM-IV generalized anxiety disorder and options for DSM-V. Depress Anxiety 27: 134–147. doi: 10.1002/da.20658

58. Ironson G, O'Cleirigh C, Fletcher MA, Laurenceau JP, Balbin E, et al. (2005) Psychosocial factors predict CD4 and viral load change in men and women with human immunodeficiency virus in the era of highly active antiretroviral treatment. Psychosom Med 67: 1013–1021. doi: 10.1097/01.psy.0000188569.58998.c8

59. Pence BW, Miller WC, Gaynes BN, Eron JJ Jr (2007) Psychiatric illness and virologic response in patients initiating highly active antiretroviral therapy. J Acquir Immune Defic Syndr 44: 159–166. doi: 10.1097/qai.0b013e31802c2f51

60. Parienti JJ, Massari V, Descamps D, Vabret A, Bouvet E, et al. (2004) Predictors of virologic failure and resistance in HIV-infected patients treated with nevirapine- or efavirenz-based antiretroviral therapy. Clin Infect Dis 38: 1311–1316. doi: 10.1086/383572

61. Ickovics JR, Hamburger ME, Vlahov D, Schoenbaum EE, Schuman P, et al. (2001) Mortality, CD4 cell count decline, and depressive symptoms among HIV-seropositive women: longitudinal analysis from the HIV Epidemiology Research Study. JAMA 285: 1466–1474. doi: 10.1001/jama.285.11.1466

62. Cohen MH, French AL, Benning L, Kovacs A, Anastos K, et al. (2002) Causes of death among women with human immunodeficiency virus infection in the era of combination antiretroviral therapy. Am J Med 113: 91–98. doi: 10.1016/s0002-9343(02)01169-5

63. Leserman J, Pence BW, Whetten K, Mugavero MJ, Thielman NM, et al. (2007) Relation of lifetime trauma and depressive symptoms to mortality in HIV. Am J Psychiatry 164: 1707–1713. doi: 10.1176/appi.ajp.2007.06111775

64. Mayne TJ, Vittinghoff E, Chesney MA, Barrett DC, Coates TJ (1996) Depressive affect and survival among gay and bisexual men infected with HIV. Arch Intern Med 156: 2233–2238. doi: 10.1001/archinte.1996.00440180095012

65. Kelly B, Raphael B, Judd F, Perdices M, Kernutt G, et al. (1998) Suicidal ideation, suicide attempts, and HIV infection. Psychosomatics 39: 405–415. doi: 10.1016/s0033-3182(98)71299-x

66. Kinsler JJ, Wong MD, Sayles JN, Davis C, Cunningham WE (2007) The effect of perceived stigma from a healthcare provider on access to care among a low-income HIV-positive population. AIDS Patient Care STDS 21: 584–592. doi: 10.1089/apc.2006.0202

67. Rao D, Kekwaletswe TC, Hosek S, Martinez J, Rodriguez F (2007) Stigma and social barriers to medication adherence with urban youth living with HIV. AIDS Care 19: 28–33. doi: 10.1080/09540120600652303

68. Ware NC, Wyatt MA, Tugenberg T (2006) Social relationships, stigma and adherence to antiretroviral therapy for HIV/AIDS. AIDS Care 18: 904–910. doi: 10.1080/09540120500330554

69. Rintamaki LS, Davis TC, Skripkauskas S, Bennett CL, Wolf MS (2006) Social stigma concerns and HIV medication adherence. AIDS Patient Care STDS 20: 359–368. doi: 10.1089/apc.2006.20.359

70. Nurutdinova D, Chrusciel T, Zeringue A, Scherrer JF, Al-Aly Z, et al. (2012) Mental health disorders and the risk of AIDS-defining illness and death in HIV-infected veterans. AIDS 26: 229–234. doi: 10.1097/qad.0b013e32834e1404

71. United States Department of Agriculture (USDA) Household Food Security Survey Module (HFSSM). U.S. Adult Food Security Survey Module. Available: http://www.ers.usda.gov/topics/food-nutrition-assistance/food-security-in-the-us/survey-tools.aspx#adult. Accessed 2012 Dec 17.

72. Statistics Canada Canadian Community Health Survey (CCHS), Cycle 2.2, Nutrition (2004): Income-Related Household Food Security in Canada. Available: http://www.hc-sc.gc.ca/fn-an/surveill/nutrition/commun/income_food_sec-sec_alim-eng.php#metho24. Accessed 2012 Dec 17.

73. Anema A, Weiser SD, Fernandes KA, Ding E, Brandson EK, et al. (2009) High prevalence of food insecurity among HIV-infected individuals receiving HAART in a resource-rich setting. AIDS Care 23: 221–230. doi: 10.1080/09540121.2010.498908

74. Vogenthaler NS, Hadley C, Lewis SJ, Rodriguez AE, Metsch LR, et al. (2010) Food insufficiency among HIV-infected crack-cocaine users in Atlanta and Miami. Public Health Nutr 13: 1478–1484. doi: 10.1017/s1368980009993181

75. Sahni S, Forrester JE, Tucker KL (2007) Assessing Dietary Intake of Drug-Abusing Hispanic Adults with and without Human Immunodeficiency Virus Infection. J Am Diet Assoc 107: 968–976. doi: 10.1016/j.jada.2007.04.003

76. Smit E, Tang A (2000) Nutritional assessment in intravenous drug users with HIV/AIDS. J Acquir Immune Defic Syndr 25: S62–69. doi: 10.1097/00042560-200010001-00010

77. Gorber SC, Tremblay M, Moher D, Gorber B (2007) A comparison of direct vs. self-report measures for assessing height, weight and body mass index: a systematic review. Obesity Reviews 8: 307–326. doi: 10.1111/j.1467-789x.2007.00347.x

Shamba Maisha: Pilot Agricultural Intervention for Food Security and HIV Health Outcomes in Kenya: Design, Methods, Baseline Results and Process Evaluation of a Cluster-Randomized Controlled Trial

Craig R. Cohen, Rachel L. Steinfeld, Elly Weke,
Elizabeth A. Bukusi, Abigail M. Hatcher,
Stephen Shiboski, Richard Rheingans,
Kate M. Scow, Lisa M. Butler, Phelgona Otieno,
Shari L. Dworkin, and Sheri D. Weiser

8.1 BACKGROUND

Despite major advances in care and treatment of those living with HIV, morbidity and mortality among people living with HIV/AIDS (PLHIV) remains unacceptably high in sub-Saharan Africa, largely due to parallel epidemics of poverty and food insecurity (Weiser et al. 2011). There are an estimated 35.3 million

© *Cohen et al.; licensee Springer. 2015; SpringerPlus, 2015, 4:122; DOI: 10.1186/s40064-015-0886-x. Distributed under the terms of the Creative Commons Attribution License (http://creativecommons.org/licenses/by/4.0/).*

PLHIV worldwide, 70.8% of whom live in sub-Saharan Africa (Global report 2013). Food insecurity, defined as "the limited or uncertain availability of nutritionally adequate, safe foods or the inability to acquire personally acceptable foods in socially acceptable ways," (Normen et al. 2005) is also highly prevalent in sub-Saharan Africa, where 240 million persons, or one in every four people, are estimated to be food insecure (FAO 2010). The prevalence of food insecurity is even higher among PLHIV in sub-Saharan Africa. Studies from Kenya and Uganda have shown that over 50% of PLHIV are moderately or severely food insecure (Mbugua et al. 2008; Weiser et al. 2010a).

Food insecurity and HIV/AIDS are leading causes of morbidity and mortality in sub-Saharan Africa and are inextricably linked, with each condition heightening vulnerability to, and worsening the severity of the other condition (Weiser et al. 2011). Food insecurity enhances HIV acquisition risk through increased risky sex and also increases susceptibility to HIV among those who are exposed (Weiser et al. 2007; Campbell et al. 2009; Webb-Girard et al. 2012; Weiser et al. 2011). Among PLHIV, food insecurity inhibits antiretroviral therapy (ART) initiation, retention in care, and ART adherence (Weiser et al. 2010b; Goudge & Ngoma 2011; Nagata et al. 2012; Weiser et al. 2009a; Wang et al. 2011; McMahon et al. 2011; Weiser et al. 2009b; Weiser et al. 2012). Food insecurity has been associated with a range of adverse clinical effects among PLHIV, including declines in physical health status (Weiser et al. 2009c; Weiser et al. 2012), decreased viral suppression (Weiser et al. 2009a; Kalichman et al. 2010a), worse immunologic status (Weiser et al. 2009c; Kalichman et al. 2010b), increased incidence of serious illness (Tsai et al. 2011), and increased mortality (Weiser et al. 2009b). In turn, HIV/AIDS worsens food insecurity by eroding economic productivity (Larson et al. 2008; McIntyre et al. 2006; Russell 2004), reducing social support due to HIV stigma (Tsai et al. 2011), and increasing medical expenses (McIntyre et al. 2006).

Although there has been a substantial increase in the allocation of international resources towards HIV care and treatment programs in Africa, food insecurity can significantly compromise the effectiveness of these programs due to its effects on morbidity and mortality as described above (Mamlin et al. 2009). As a result, the World Health Organization, UNAIDS and the World Food Programme have recommended integrating sustainable food production strategies into HIV/AIDS programming (Food and Nutrition Technical Assistance. HIV/AIDS: A Guide For Nutritional Care and Support; Nutrition and HIV/AIDS 2001; Blumberg & Dickey 2003; World Food Program 2003). Specifically, UNAIDS calls for international partners to "fund multisectoral HIV programming that incorporates effective food and nutrition interventions, in line with scale-up towards universal access to prevention, treatment, care and support" (UNAIDS Policy Brief 2008). Yet, little research exists to document the beneficial effect of food security or sustainable agricultural interventions on antiretroviral (ARV) adherence, HIV clinical outcomes, women's empowerment and HIV transmission risk behaviors among PLHIV in Africa or elsewhere.

To date, no randomized controlled trials have been conducted in resource-limited settings to examine the impacts of either food supplementation or sustainable food production strategies on HIV morbidity and mortality (Mahlungulu et al. 2007). Several small studies in developing countries have demonstrated the potential for programs that address food security to affect health outcomes among PLHIV (Mamlin et al. 2009; Cantrell et al. 2008; Agricultural initiatives for health in Haiti; Ochai 2008; Njenga et al. 2009; Byron et al. 2008). Yet, existing intervention approaches to impacting food security have focused primarily on direct macronutrient supplementation, which may be somewhat limited in its scalability and sustainability (Sztam et al. 2010). Livelihood interventions, which address upstream causes of food insecurity, may have a better chance of improving health outcomes, and may be more sustainable. Similarly, while microcredit programs can improve health and prevent disease acquisition by targeting poverty and gender inequality (Ashburn et al. 2008; Kim et al. 2008; Schuler & Hashemi 1994), they have been criticized in terms of their effectiveness as a stand-alone strategy. As a result, experts have recommended integrating microfinance and other livelihood approaches to maximize HIV prevention and treatment efforts and reduce poverty (Dworkin & Blankenship 2009; Weinhardt et al. 2009). Income generating activities are well suited to improving food security (Diagne 1998; Doocy et al. 2005), and to retaining patients in HIV care (Mamlin et al. 2009; Gomez et al. 2004).

Following our previously published theoretical model (Weiser et al. 2011), we set out to test the impact of a multisectoral agricultural intervention on HIV health outcomes and transmission risk behaviors in rural Kenya. In a previous small feasibility study conducted by our group, we showed that using a human powered irrigation hip pump combined with a microfinance loan led to increases in crop yields, household income, CD4 counts and BMI (Pandit et al. 2010). In the current study, we developed and tested a modified version of this combination agricultural and microfinance intervention called Shamba Maisha, Kiswhahili for "farm life," in a small community randomized control trial. In this study, we aimed to explore the acceptability and feasibility of the intervention and control conditions that will be used in a subsequent, larger cluster randomized controlled trial, and examined preliminary impacts on outcomes of interest. Two similar district hospitals in southern Nyanza Province were randomized: one to the intervention and the other to the control group. We hypothesized that this multisectoral intervention will improve food insecurity, household wealth and HIV health outcomes. We developed and tested a theoretical framework for the pathways through which this multisectoral agricultural intervention may improve health. In this paper, we described our conceptual framework, study methods, baseline findings, as well as process evaluation findings of successes and challenges with implementation.

8.2 METHODS

8.2.1 Setting

The study took place in Rongo and Migori districts in Nyanza Province, Kenya. As of 2008, the HIV prevalence in Nyanza Province, Kenya, was estimated to be 15.3%, more than twice the national average (Kenya National Bureau of Statistics and ICF Macro 2010). Nearly all HIV-affected households in a recent Kenyan survey were considered to be moderately or severely food insecure (Mbugua et al. 2008). The province also has a significant shortage of accessible water making communities vulnerable to the impacts of drought, and a heavy dependence on an unstable agricultural sector. Farming and fishing are the primary means of income generation in Nyanza Province. Lack of irrigation and unpredictable rainfall leading to an inconsistent water supply remains a central barrier to successful farming for many in the region (Government of Kenya 2008).

8.2.2 Description of Intervention

The *Shamba Maisha* intervention consisted of three components: 1) a microfinance loan (~$150) to purchase the farming commodities, 2) a micro-irrigation pump, seeds, fertilizer and pesticides, and 3) trainings in sustainable agricultural practices and financial literacy. KickStart, an international non-governmental organization (NGO), along with technical support of agricultural experts from the University of California Davis, led the agricultural component of the intervention. *Adok Timo*, a microfinance institution in Nyanza Province, implemented the economic aspects of the intervention with support from the UCSF and KEMRI research team. The intervention components are described below.

8.2.3 Loan Program

The microfinance loans were managed by *Adok Timo,* which has branches in Nyanza Province. The intervention group received training on financial management and marketing skills prior to receiving the loan. All trainings took place on participant's farms or a nearby location. Participants were required to save 500 Kenyan shillings (~$6.00 USD) prior to receiving the loan. Each participant received vouchers to purchase the following items: the Hip Pump, 50 feet of hosing, fertilizer, and government certified seeds. These materials were purchased at a local farm store ("agrovet"). Loan repayment could begin at any time, but farmers were expected to make a minimum payment after the first harvest, usually 4–6 months after planting. Farmers were expected to repay in full by the end of two harvest seasons (approximately one year); however, if regular payments have been made, this deadline was extended based on the guidelines set out by the microfinance institution. Participants were not asked to forfeit personal belongings to cover loan payments. Control participants were eligible for the microcredit loan and the *Kickstart* Hip Pump at the end of the 1-year follow-up period.

KickStart water pump and agricultural training: Irrigation technologies have been shown to improve agricultural output. (About KickStart) Recognizing the need for improved agricultural tools for poor farmers in Kenya, Kickstart developed a low-cost irrigation pump. These pumps enable farmers to irrigate their crops year-round avoiding dependence on seasonal rainfall thus capitalizing on higher crop prices in the marketplace. In prior evaluations in Kisumu Kenya, farmers using this hip pump have been able to enjoy up to a ten-fold increase in income (Brandsma 2003; Kihia & Kamau 1999; Stevens 2002; World Bank 2010).

Prior to receiving the microfinance loan, participants in the intervention group received eight training modules. The training modules were delivered to groups of farmers on participant's farms or at a nearby location by agricultural trainers from Kickstart, the loan officer from *Adok Timo,* and the study coordinator. The agricultural portion of the training included a didactic session and practical demonstrations on sustainable farming techniques, seed selection, soil and water conservation, fertilization and crop rotation, integrated pest and disease management (IPM), pre & post-harvest handling and marketing, the use of the MoneyMaker Hip Pump, and identifying improved market access for selling horticultural products. The field-based practical trainings were based on a training needs assessment and the crops selected by the participants. These trainings focused on plant spacing, seed selection, seed rate, crop combination, and a cost-benefit analysis for each crop selected and IPM. Participants were particularly interested in locally available materials for IPM like wood ash, Mexican marigold and livestock urine in controlling pests on crops to enable them to reduce cost and maximize profits. Participants also conducted a market survey as part of the pre & post-harvest handling and marketing which revealed a large demand in the local markets for green vegetables and watermelon. The intervention participants also had an in-depth training on financial record keeping, the importance on savings and loans, and the basics of group dynamics.

8.2.4   Intervention Model

The primary aim of the pilot study was to test if a multisectoral agricultural intervention improves food security and HIV health outcomes. We hypothesized that the proposed multisectoral agricultural intervention will improve food security, prevent treatment failure, reduce co-morbidities, and decrease secondary HIV transmission risk among PLHIV.

We developed an evidenced-based causal framework (Figure 1) to understand the pathways by which food insecurity negatively impacts health outcomes. Food insecurity negatively impacts health outcomes through nutritional, behavioral, and mental health pathways, which emerge directly from the definition of food insecurity (Weiser et al. 2011). In terms of nutritional pathways, food insecurity has been associated with macronutrient and micronutrient malnutrition (Rose & Oliveira 1997; Lee & Frongillo 2001), and weight loss, low body mass index (BMI), and low albumin have been shown to hasten progression to AIDS and death (Stringer et al. 2006; Zachariah et al. 2006; Johannessen et al. 2008). HIV also increases metabolic requirements (Babameto & Kotler 1997; Macallan et al. 1995) and is associated with diarrhea and malabsorption of fat and carbohydrates (Babameto & Kotler 1997; Kotler et al. 1990; Stack et al. 1996; Fields-Gardner & Fergusson 2004). Lack of food may also impede optimal absorption of certain ARVs (Gustavson et al. 2000; Bardsley-Elliot & Plosker 2000; Food and Agriculture Organization 2010), which may contribute to treatment failure. In terms of behavioral pathways: Food insecurity is an important cause of ART non-adherence and treatment interruptions, (Weiser et al. 2010b; Weiser et al. 2009a; Kalichman et al. 2010a; Weiser et al. 2009d; Tuller et al. 2010) which are both well-known determinants of HIV treatment outcomes (Parienti et al. 2004; Parienti et al. 2008; Oyugi et al. 2007). In addition to ART non-adherence and treatment interruptions, as a result of competing demands between food and other resources, food insecure individuals often miss scheduled clinic visits, and may be less likely to initiate ART (Weiser et al. 2010b; Tuller et al. 2010; Weiser et al. 2012). In terms of mental health pathways, food insecurity has been associated with depression and poor mental health status in sub-Saharan Africa and elsewhere, (Weiser et al. 2009c; Tsai et al. 2011; Anema et al. 2011) which in turn have been shown to independently contribute to lower ART adherence and worse HIV clinical outcomes (Ickovics et al. 2001; Tucker et al. 2003; Evans et al. 2002; Weiser et al. 2004). Food insecurity and poverty also contribute to lower levels of empowerment, including women's empowerment, which can negatively impact health outcomes for HIV. For example, in Uganda, among HIV-infected women low sexual relationship power is associated with malnutrition (Siedner et al. 2012), depression (Hatcher et al. 2012), and worse virologic outcomes (Weiser et al. 2010a, b).

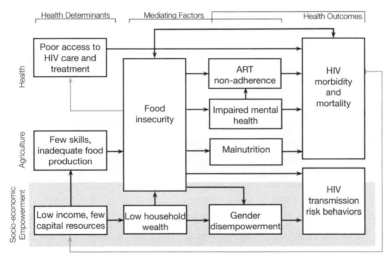

FIGURE 8.1   *Shamba Maisha* causal framework

Drawing upon this framework, we developed the intervention framework to describe the pathway by which we believe the multisectoral agrigultural intervention impacts on mediating factors and eventually improves health outcomes among HIV-infected adults (Figure 2). Specifically, we hypothesized that: 1) The three intervention components together would directly lead to improvements in food security and household wealth (most proximal mediators); 2) Changes in food security and household wealth would, in turn, contribute to less macronutrient and micronutrient malnutrition (nutritional pathway), less anxiety, stress and depression (mental health pathway), improved ART adherence and retention in care (behavioral pathway) and improved gender empowerment; and 3) Through these pathways, the intervention would ultimately contribute to decreased HIV morbidity and mortality and fewer HIV transmission risk behaviors.

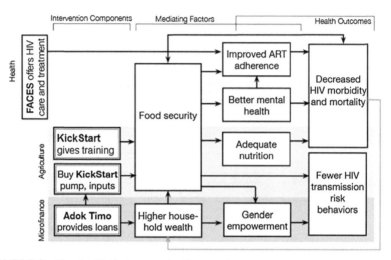

FIGURE 8.2   *Shamba Maisha* intervention framework

8.2.5   Randomization

We used a two-step process to select two sites from among 68 government health facilities in Rongo, Migori and Nyatike districts supported by Family AIDS Care & Education Services (FACES), a collaboration between the University of California, San Francisco (UCSF), and the Kenyan Medical Research Institute (KEMRI) (Lewis Kulzer et al. 2012). First, we limited to 20 sites where there were an adequate number of PLHIV on ART at the time of the study, and where farming was a primary means of livelihood in the communities. Next, we selected two government health facilities for the pilot randomized controlled trial (RCT) according to the following criteria: a) having adequate numbers of PLHIV on ART with non-overlapping catchment areas; b) having catchment areas that were relatively similar to one another according to rainfall patterns, health, topography and socioeconomic status; and c) meeting practical requirements for the intervention, including acceptable characteristics in topography to enable use of the water pump, having water access year round, and having favorable soil composition to potentiate the intervention. One site was randomly selected as the intervention site and the other the control site by the study's biostatistician who used a computer random number generator and was not involved in fieldwork. Clinics, healthcare providers, patients, and researchers involved in implementing the study were not blinded to the allocation.

8.2.6 Ethics Statement

The study was approved by the Committee on Human Research at UCSF and the Ethical Review Committee at KEMRI. All participants in the study gave written informed consent prior to enrollment in the study, and had the cost of their transportation reimbursed up to 800 Ksh per clinic-based interviews (~$9.4 USD) and 400 Ksh for home-based interviews (~$4.70). This clinical trial was registered at ClinicalTrials.gov (NCT01548599).

8.2.7 Participants and Recruitment

The study population includes HIV-infected patients between the ages of 18–49 years receiving ART, have access to farm land and surface water, and have demonstrated evidence of moderate to severe food insecurity or malnutrition during the year preceding the study. Participants were recruited from Rongo (intervention site) and Migori (control site) District Hospitals in Nyanza Province beginning in April 2012. Research assistants introduced the study to patients waiting to be seen at the HIV clinic at the two health facilities. Individuals who expressed interest were consented and screened for eligibility. Among interested and potentially eligible individuals, home visits were conducted to verify that the participant had access to farming land and surface water. In addition to the eligibility criteria above, the study also aimed to ensure that at least 40% of participants at each site were of each gender. Individuals who met eligibility criteria were enrolled in the study after providing written informed consent. At intervention sites, participants were enrolled in a savings program in anticipation of receiving the asset loan. Control participants also agreed to save the down payment (~$6) required for the loan by the study end. A total of 140 HIV-positive individuals were enrolled in the study during the period of April – July 2011 (See Figure 3); an additional two (1.4%) eligible screened participants declined to participate in the study.

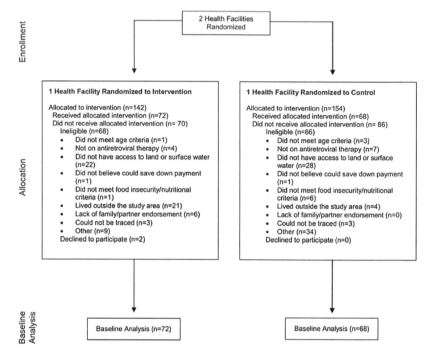

FIGURE 8.3 *Shamba Maisha* consort diagram

8.2.8 Data Collection

Participants were followed quarterly for structured interviews and ART adherence assessments by unannounced pill counts at their home. Anthropometry (i.e., mid upper arm circumference measurements [MUAC] and BMI measurements) and phlebotomy for viral load and CD4 determinations were conducted every six months. Viral load testing was performed on venous blood at the Centers for Disease Control-Kenya laboratory on the COBAS TaqMan HIV viral load platform (Roche Molecular Diagnostics, Pleasanton, CA) with a lower limit of detection of <20 copies/mL. Absolute CD4 count testing was performed on whole blood using the BD FACSCount (BD Bioscience, San Jose, CA). In addition, clinical data were abstracted from participant's medical records.

Structured interviews for clinical and sensitive behavioral data were administered at baseline, 6 months and 12 months at the clinic and included the following data (see Table 1): health care access and competing demands, ARV history and adherence, physical and mental health, social support, women's empowerment,

TABLE 8.1 Measurements

Path	Measurement	Definition	Frequency
Nutritional Pathway	Nutritional status	Nutritional status was assessed through body mass index (BMI) and mid-upper arm circumference (MUAC), commonly used to assess nutritional status (Physical Status 1995; Collins et al. 2000). The BMI reflects protein and fat reserves (James et al. 1988) and was assessed using an established grading system (Ferro-Luzzi et al. 1992). For MUAC, we used WHO sex-specific cut-offs of 22.0 cm for females and 23.0 cm for males with chronic energy deficiency (Doocy et al. 2005).	Semi-annually
	Food frequency	Food frequency, the number of different foods or food groups and the frequency consumed over a given reference period (Hoddinott & Yohannes 2002), as adapted from the World Food Programme Food Consumption Score was collected.	Quarterly
Behavioral Pathway	Pill count ART adherence	Participants received an unannounced visit to inventory medications and count pills (Bangsberg et al. 2001a; Bangsberg et al. 2001b; Bangsberg et al. 2000), a technique closely correlated with electronically monitored adherence, HIV viral load (Bangsberg et al. 2000) and progression to AIDS (Bangsberg et al. 2001c). The count of existing pills was reconciled with the participant's pharmacy refill history to determine the percentage of pills not yet consumed.	Quarterly
	Competing demands	Questions were modified from Gelberg and Anderson's Behavioral Model for Vulnerable Populations (Gelberg et al. 2000; Gelberg et al. 1997) to assess how often lack of food interferes with ability to procure drugs or visit the clinic	Semi-annually
	Healthcare access	Utilization of health care services including hospitalizations and clinic visits over the preceding 6 months were collected	Semi-annually
Mental Health Pathway	Mental health & depression	Mental health status was measured using the Medical Outcomes Study HIV Health Survey (MOS-HIV), a tool for assessing health-related quality of life (Wu et al. 1991) that has been validated in resource-limited settings (Chatterton et al. 1999; Mast et al. 2004). Depression was screened using the Hopkins Symptom Check-list for depression, a 15-item scale (Derogatis et al. 1974) which has been validated in sub-Saharan Africa (Bolton et al. 2004).	Semi-annually
	HIV-related stigma	We used the Internalized AIDS-Related Stigma Scale (Kalichman et al. 2009).	Semi-annually

TABLE 8.1 (*Continued*)

Path	Measurement	Definition	Frequency
	Disclosure of HIV status	We asked about disclosure of HIV status to partners, family members, friends, colleagues, and public. These questions were adapted from our previous studies in Uganda, Botswana and Swaziland (Weiser et al. 2007; Wolfe et al. 2008; Tsai et al. 2013).	Semi-annually
	Alcohol use	To measure alcohol use, we adapted the Alcohol Use Disorders Identification Test (AUDIT-C) indicators. The AUDIT-C is a 3-item alcohol screen that can help identify persons who are hazardous drinkers or have active alcohol use disorders.	Semi-annually
Empower-ment	Gender empowerment	Empowerment indicators were adapted from a large cluster-randomized trial of an intervention including: greater challenges to established gender roles, communication with relationship partner about sexual matters in the prior 3 months, measures of financial decision-making, measures of attitudes towards gender roles and gender-based violence, and experience of controlling behavior by relationship partner in prior 3 months (Pronyk et al. 2006). In addition, we used the Sexual Relationship Power Scale (SRPS) (Pulerwitz et al. 2000), which conceptualizes sexual relationship power as a multi-dimensional construct consisting of relationship control and decision making dominance. The SRPS has been used successfully in observational research conducted in South Africa (Dunkle et al. 2004; Jewkes et al. 2010) and Uganda.(Weiser et al. 2010a, b) We also collected data on sexual victimization and perpetration in the prior 3 months.	Semi-annually
Proximal Mediators	Food insecurity	The Household Food Insecurity Access Scale (HFIAS) has been validated in eight countries (Coates et al. 2006a; Swindale & Bilinsky 2006; Frongillo & Nanama 2006; Coates et al. 2006b) and used successfully by our team in rural Uganda. (Weiser et al. 2010a; Tsai et al. 2011; Weiser et al. 2012; Tsai et al. 2012; Weiser et al. 2010a, b; Miller et al. 2011; Tsai et al. 2011)	Semi-annually
	Agricultural measures	Agricultural measures were adapted from outcome evaluations developed by Kickstart in Kenya and supplemented by outcome indicators found in an earlier pilot study and a rural assessment. These measures were designed to evaluate uptake and adoption, and to measure changes in agricultural practices including crop diversity and agricultural practices and production. In addition, we evaluated the effectiveness of the training, and specific topics within, so as to refine the training for the subsequent larger cluster-randomized trial.	Quarterly

TABLE 8.1 (*Continued*)

Path	Measurement	Definition	Frequency
Behavioral Outcome	Household economic indicators	A modification of the World Bank Living Standards Measurement Study (LSMS) questionnaire (Grosh & Glewwe 1998) was used to measure: a) expenditures (food, health, education and productive investments); b) consumption (food and non-food); c) income (from agriculture and all sources); and d) inter-household commodity and cash transfers.	Quarterly
	Risky sexual behaviors	The primary transmission risk outcome was unprotected sex. Other outcomes included: number of non-spousal/non-cohabiting sexual partners, sex-exchange (exchanging sex for money, food, or other resources) Dupas & Robinson 2010); Robinson & Yeh 2011).	Semi-annually
Health Outcomes	HIV-related mortality	Burial permits and information from family members were used to determine cause of death.	As needed
	Viral load suppression	Viral load testing was performed on venous blood on the COBAS TaqMan HIV viral load platform (Roche Molecular Diagnostics, Pleasanton, CA) with a lower limit of detection of <20 copies/mL.	Semi-annually
	CD4 Count	We abstracted data for CD4 counts from participant's medical records. Absolute CD4 count testing was performed on whole blood using the BD FACSCount (BD Bioscience, San Jose, CA).	Semi-annually
	HIV morbidity	HIV morbidity was measured through key outcomes from the medical record. We abstracted data every 3 months for ART treatment interruptions and episodes of opportunistic infections. We also gathered self-report data on opportunistic infections and symptoms during structured interviews.	Quarterly/Semi-annually
	Physical health	Health status was measured using the MOS-HIV, a tool for assessing health-related quality of life (Wu et al. 1991) that has been validated in resource-limited settings (Chatterton et al. 1999.; Mast et al. 2004).	Semi-annually
Covariates	Demographics	Age, religion, education, marital/partnership status, number of children and household census.	Baseline

TABLE 8.1 *(Continued)*

Path	Measurement	Definition	Frequency
	Social support	To measure social support, we adapted the Functional Social Support Scale (Antelman et al. 2001), a modified version of the Duke University-University of North Carolina Functional Support Questionnaire (Broadhead et al. 1988) consisting of questions related to perceived emotional and instrumental support. Higher scores reflect higher levels of social support.	Semi-annually
	ART history and Self-reported ART adherence	Detailed ART history, ART self-reported adherence and barriers to ART adherence were collected. For self-report adherence, we used the visual analog scale (Oyugi et al. 2004) and the three day recall.	Semi-annually

stigma, HIV disclosure, alcohol use, and sexual behavior. Additionally, quarterly home visits were conducted to collect unannounced pill counts for ART adherence, food frequency, food security, income, housing, assets and wealth, and specific data related to agricultural output, labor, harvesting, marketing, and irrigation.

8.2.9 Data Collection Methods

Structured interviews administered at the health facility and at participants' farms or homes were conducted by trained research assistants who used a handheld computer tablet for data entry (Morotola™ Xoom Android Tablets operating Open Data Kit (ODK) Collect (Hiarlaithe et al. 2014). Research assistants conducted structured interviews in the local language (Dholuo) or English and completed paper-based data collection forms for the majority of baseline data collection (Table 1). Forms were later entered into the tablet using ODK Collect. Research assistants entered subsequent data directly into the tablet. Data for anthropometry (MUAC and BMI), and viral load results were entered into ODK Collect forms (Table 1). Medical history and CD4 determinations were abstracted from patient records and entered into the tablet (Table 1).

8.2.10 Monitoring

The study team conducted internal monitoring such as form completion and quality control of data entry monthly. Procedures to promote data quality including range and logical checks were built into the data entry program, and we ran a series of additional error checks on the databases following data entry. In addition, we used a rainfall gauge at the two study sites to record total precipitation by month.

8.2.11 Process Evaluation Methods

In order to understand the implementation successes and challenges, we conducted a detailed process evaluation alongside the Shamba Maisha trial. We interviewed 40 intervention participants and 20 key informants at two time points (3–5 months after study start and study end) to understand successes and challenges with implementation (for a total of n = 120 interviews).

Interviews were conducted in English, Dhuluo or Kiswahili, and lasted between 45 minutes and 2 hours. We used a semi-structured interview guides to probe for feedback on recruitment and retention as well the key intervention components (the microcredit loan, the agricultural training, and the use of the hip pump), and elicit suggestions on how to improve each of these components. Interviews were transcribed verbatim and (as necessary) translated into English. Data were managed using Dedoose, (Dedoose Version 5.0.11 et al. 2014) a qualitative software that allows for real-time access to a secured database by a number of people in an analysis team. In line with previous qualitative process evaluations of complex interventions (Hargreaves et al. 2010), we used an inductive-deductive approach. First interviews were coded for broad themes by four researchers, using a structured coding framework developed from topics covered in the interview guide. Next, a second stage of inductive coding allowed sub-themes to emerge. Process evaluation findings are used to suggest how these might help refine the intervention in a future trial.

8.2.12   Statistical Methods

For this baseline manuscript, we compared baseline characteristics between participants enrolled in the intervention and control sites using proportions for categorical variables and means or medians (as appropriate) for continuous variables. An assessment of the distribution (normally or skewed) of continuous variables was performed. Statistical tests for descriptive analyses included chi-square tests for categorical variables, and Wilcoxon rank-sum or t-tests and for continuous variables.

8.2.13   Analysis of Outcome Data

This pilot was not powered for formal significance testing of the intervention effect on primary health, behavioral and gender empowerment outcomes. Rather, analyses will focus on assessing the effects of the intervention on proximal mediating factors (Figure 2), and separately on primary outcomes. Analyses for mediating variables will generally treat scores as continuous measures. For example, food security will be measured using the Household Food Insecurity Access Scale (HFAIS) score at the 1-year visit; change in household economic indicators over the 1-year follow-up period will be measured by subtracting baseline from 1-year follow-up data for household expenditures, consumption and income (from agricultural and other sources). Initial comparisons will be based on group-specific descriptive summaries of observed outcomes and rank-based tests comparing outcomes between groups. We will also use regression methods to compare outcomes between groups and control for baseline characteristics. Analyses for primary health outcomes will proceed similarly, with appropriate choices of model for outcome type. For example, we will use pooled logistic regression for between-group comparisons of rates of viral load suppression, and mixed effects regression to compare changes in CD4 counts between groups. We will also make preliminary assessments of degree of mediation in models for primary outcomes via inclusion of mediating factors, with assessment of direct and indirect intervention effects of key mediating variables (Petersen et al. 2006). Finally, we will also perform intent-to-treat and per-protocol analyses of the pilot data, with the per-protocol subset limited to participants that obtained and used the intervention consistently over the 12-months of follow-up. Although these analyses will likely be underpowered for formal testing purposes, the resulting estimates and confidence intervals will be important for planning for the subsequent RCT.

8.3   RESULTS

8.3.1   Screening and Enrollment

Figure 3 describes study screening and enrollment numbers, along with reasons for participant ineligibility. We screened 142 and 154 adults, respectively at the intervention and control sites. We enrolled 72 and 68 participants at the intervention and control sites, respectively; four participants withdrew from the study, four from the intervention site and zero from the control site. One participant in the intervention arm failed to save the down payment necessary to receive the loan and was subsequently withdrawn from the study. The most common reasons for study ineligibility of screened participants included: did not have access to land or surface water, lived outside the study area, not on ART, and "other" reasons.

8.3.2   Baseline Characteristics of Participants

At baseline (Table 2) participants at the intervention and control sites, respectively were similar in age (37 vs. 38 years), gender (51% female in both), education (19% vs. 18% completed secondary school) and marital status (75% vs. 79% married). Mental health and gender empowerment measures were similar among participants enrolled at the intervention and control sites. Reported monthly household income was lower among participants enrolled at the intervention in comparison to the control site (~$104) vs. (~$273), p = 0.008).

TABLE 8.2 Comparison of baseline sociodemographic, economic, sexual risk behavior, dietary intake, clinical, mental health, gender empowerment, clinical and laboratory findings of participants enrolled in *Shamba Maisha* at the intervention and control sites

	Intervention N = 72	Control N = 68	P-value
Sociodemographic variables	n (%)	n (%)	
Mean age in years (SE)	37 (0.80)	38 (0.80)	0.16
Female gender	37 (51)	35 (51)	0.92
Education ≥ secondary school	14 (19)	12 (18)	0.82
Current married	54 (75)	54 (79)	0.25
Mean number of people in household (SE)	6 (0.2)	7 (0.3)	0.12
Food security level			
Moderately food insecure	14 (20)	14 (21)	0.86
Severely food insecure	57 (80)	53 (78)	0.86
Body Mass Index (BMI) <18.5	13 (18)	5 (7)	0.054
Economic indicators			
Mean (SE) Monthly Household Income (USD*)	104 (21.3)	273 (53.1)	0.008
Land ownership (Self or Jointly)	60 (83)	49 (73)	0.14
Sexual risk behavior			
Any unprotected sex	13 (18)	12 (18)	0.95
Sex exchange (ever)	15 (21)	28 (41)	0.009
Sex exchange (last 3 months)	2 (3)	6 (9)	0.12
>1 Sexual partners in the past 3 months	9 (13)	9 (13)	0.92
Dietary intake (frequency of consumption - times/week)			
Grains mean (SE)	19.0 (0.8)	20.8(0.8)	0.1320
Vege`s mean (SE)	20.3 (0.9)	25.9 (0.7)	<0.0001
Fruit mean (SE)	9.0 (0.6)	10.6 (0.9)	0.1352
Meat, Poultry mean (SE)	3.4 (0.3)	5.3 (0.3)	<0.00001
Eggs mean (SE)	0.9 (0.1)	1.3 (0.2)	0.117
Dairy mean (SE)	1.9 (0.3)	4.5 (0.5)	<0.0001

TABLE 8.2   Comparison

	Intervention N = 72	Control N = 68	P-value
Cooking fat mean (SE)	5.0 (0.3)	6.8 (0.1)	<0.00001
Sweets mean (SE)	7.1 (0.7)	11.9 (0.7)	<0.00001
Mental health			
Depression score ≥1.75 (clinical depression)	1 (1)	4 (6)	0.2
Mental health summary score (SE)	57.1 (3.4)	42.8 (5.6)	0.15
Mean internalized AIDS related stigma scale (SE)	14.1 (0.2)	13.7 (0.2)	0.19
Disclosed to husband/wife/partner	65 (97)	63 (95)	0.68
Mean Social Support Score (SE)	53.4 (10.8)	35.8 (8.3)	0.2
Problem drinking based on audit C	1 (1)	2 (3)	0.53
Gender empowerment			
Mean sexual relationship power scale (women only) (SE)	54 (10)	47 (8)	0.59
Mean decision making dominance scale (SE)	46 (5)	42 (4)	0.46
Clinical, Laboratory and antiretroviral (ART) adherence			
WHO clinical stage 3 or 4	15 (21)	22 (33)	0.12
Median CD4 (IQR) cells/mm^3	446 (330–629)	475 (356–679)	0.33
Baseline CD4 ≤ 350 cells/mm^3	20 (29)	16 (24)	0.467
Viral load above the limit of detection (≥20 copies/mL)	34 (49)	19 (28)	0.010
Physical health summary score (SE)	54.3 (3.6)	45.5 (4.0)	0.03
Hospitalized in the last three months	3 (4)	5 (7)	0.4
Reported ART adherence per Visual Analogue Scale < 90%	0 (0)	6 (9)	0.01

*Converted Kenyan shillings (Ksh) to US dollar (85 Ksh/$).

A similar proportion of participants at the intervention and control sites reported moderate and severe food insecurity based on the HFAIS, although a greater proportion of participants at the intervention site had a low BMI in comparison to participants at the control site (18% vs. 7%, p = 0.054) (Table 2). In regards to dietary intake, participants enrolled at the intervention site reported less frequency of vegetable, meat/poultry, dairy, cooking fat, and sweets than participants enrolled at the control site (Table 2).

Twenty-one percent and 33%, respectively of intervention and control site participants had WHO stage 3 or 4 disease. While median CD4 count was similar between arms (intervention: 446 cells/mm^3 vs. control: 475 cells/mm^3, p = 0.33), a greater proportion of participants enrolled at the intervention arm had a detectable HIV viral load than participants enrolled at the control arm (49% vs. 28%, respectively, p < 0.010); Self-reported ART adherence < 90% per visual analogue scale was rare in both arms of the trial (Table 2).

8.3.3   Implementation Successes and Challenges

A variety of implementation successes and challenges were documented during the qualitative process evaluation. Positive aspects of *Shamba Maisha* included high acceptability during recruitment, successes with agricultural and financial training, and labor savings from using the pump. Implementation challenges included considerable concern about repaying loans, agricultural and irrigation challenges related to weather patterns, and a challenging partnership with the microfinance institution.

8.3.4   Successes with Shamba Maisha Implementation

1. *Strong willingness to join.* Nearly all participants interviewed expressed a strong interest in taking part in *Shamba Maisha*. Many had expectations that this intervention would improve their households and lives as expressed by one participant:

   "What motivated me more to join the project was the training that they were coming to train us, they gave us some very good training. So when I heard about the type of training, the targets they put for us and the amount of money we would get … that if we got it then life would change for the better." (Male participant, 3 months, 31 years old).

2. *Excellent agricultural training.* The agricultural portion of the training was a success and participants noted positive impacts even before they received the agricultural inputs, including increased and diversified crop production. The newly acquired farming skills were discussed with great pride by participants. The agricultural techniques resulted in larger

harvests from smaller pieces of land – an outcome that many participants cherished:

"Before Shamba Maisha we just used to farm very huge pieces of land and get very small harvests. From the training we realized that we are able to farm a small piece of land and get a bumper harvest. This was really an eye opener for us." (Female participant, 3 months, 42 years old).

Another woman explained crucial techniques around plowing, ridge planting, and nursery beds:

"We were taught on how to farm in order to get something good; for you to realize a good harvest, you need to plough deeply, then break it into more fine particles. Then you make ridges that we were given to help us in making the seedbeds and nursery. You make the nursery then plant them then later transfer then to the main farm." (Female participant, 3 months, 45 years old).

Some participants even explained how neighbors were drawn to them as a result of the improved techniques:

Because of *Shamba Maisha,* I was able to embrace a modern way of farming that has attracted people from all over to come and learn from it. (Male participant, 12 months, 44 years old).

3. *Bolstering financial skills.* Some participants spoke of the utility of financial trainings, in particular the lessons around savings and record keeping:

    "On savings we learnt a lot from the leader on the microfinance team on how we can save in the bank, we also learnt on record keeping while farming and how we can build our lives relying on the trainings…That was very good." (Male participant, 3 months, 38 years old).

4. *Ease of irrigation with water pumps.* For a number of *Shamba Maisha* participants, the pump was easier to implement than previous irrigation strategies, such as bucket irrigation, leading to labor savings.

"I use the pump with a well. It helps in irrigation, at the times when these things need water I can use it. And I find it to be easier as compared to how I could have done it in the past where I could have taken maybe a bucket and use it to pour water." (Male participant, 3 months, 33 years old).

"Before I used to fetch water from the borehole then I used a calabash to pour water directly. But now once I get two people, I just take the machine and pump water straight from the water source and irrigate just by pressing. So I use less energy in irrigation unlike before." (Male participant, 12 months, 44 years old).

8.3.5   Barriers to Shamba Maisha Implementation

Despite these successes during implementation, participants noted a number of challenges, and had some recommendation for improving the intervention.

1. *Fear of loan repayment.* Some participants talked about the loan repayment as a potentially discouraging aspect of the intervention:

   "We were going to be given loan and we'd be expected to repay some money back. This discouraged people and it's still a factor that continue to discourage others…. here in Luo, they are scared of 'debt recoveries.'" (Male participant, 3 months, 43 years old).

2. *Lack of choice in how to use loan.* While many were satisfied with the commodities provided as part of *Shamba Maisha,* some suggested that to improve the intervention, *Shamba Maisha* should provide the loan but allow people to decide how best to use it. For example, one man spoke about preference for livestock over farming:

   "Farming, it is very difficult and requires you to take a lot of time, because people like us if you are doing it, it is something that you do full time. But we were thinking that if there was a different way other than farming, if one could choose what he could do then I think it could be better. … If I was told to choose then I would prefer to be given a cow, because that is something that I know and I have had some experience with." (Male participant, 3 months, 35 years old).

Another participant wished for crops that were less perishable so that they can be stored and sold when the market was better:

"In my own view to help people more, if we could be introduced to a new type of farming or crop, that could uplift us because vegetables are little in terms of production and also are very perishable. So if you can introduce some new crops that even after harvesting can be stored until they fetch a good price in the market. You can keep them and later decide on the quantity to sell to repay the loan." (Female, 12 months, 40 years old).

3. *Agricultural Challenges.* While agricultural productivity was reported to have improved by the majority of participants, variability in weather patterns impacted crop production throughout the course of the study period and across the communities. There were periods of flooding and hail storms which adversely affected crop production. For example, one woman complained of weather-related crop damage, despite describing having been very well trained to farm in the program "All my crops have been destroyed by the hailstorm and I have no means of repaying the loan." (Female participant, 3 months, 38 years old) Rain gauges were used to monitor precipitation at the district hospitals in the intervention and control communities. Annual rainfall was reported to be higher than normal over the course of the one-year intervention with extended rainy seasons. As a result, the irrigation technology was not utilized as much as expected, decreasing some of the potential advantages.

    Irrigation is good, though we haven't embraced it to an extent that it can fully benefit us because when we started it … the rain started. Our pumps are not being used because currently it's rainy season and you can't really sow vegetable seeds when there is too much rain. (Male participant, 12 months, 44 years old).

4. *Pump Challenges.* Although most participants reported that the pumps were easy to use, some male and female participants had difficulty operating the pumps on their own. As described by one participant:

    It is impossible to operate it comfortably as an individual. It demands that it is operated by two or three people with one person pumping the water as one directs the pipe to the shamba [farm]. When I used it, my sons had to assist me. (Female participant, 3 months, 30 years old).

5. *Challenges with Microfinance Institution.* The local microfinance organi-
    zation selected for the study had poor governance and was unable to ful-
    fill the obligations of the study. Specifically, there were significant delays
    in dispersal of the loans and fiscal insolvency, leading to the closure of
    the local branch:

> It [The microfinance organization] could be closed for even 3 months. You
> go there and find the offices are closed and sometimes your money is in
> there…so it leaves you with a lot of doubts…And that is the main reason
> why many people ended up having fears that their money would get lost
> here. (Male participant, 3 months, 28 years old).

Based on this finding, which emerged early in the process evaluation data
collection, study leadership terminated the partnership with *Adok Timo* and
transferred the loans to a local NGO which managed the loan collection for the
remainder of the study period.

8.4   DISCUSSION

This pilot study was designed to determine the acceptability and feasibility of the
intervention and control conditions, and to determine the preliminary impact of
the intervention on mediating outcomes (food security, and household economic
indicators). In the baseline evaluation of *Shamba Maisha*, we demonstrated that
screening and enrollment into the intervention and control groups was rapid.
Enrollment of the 140 participants took only four months, and the screening to
enrollment ratio was similar between study arms. Thus, patients at the two study
facilities were interested to screen and enroll in the study, i.e. enrollment into
the control conditions did not appear to mitigate interest in study participation.
This finding has important implications in regards to feasibility for advancing the
intervention into a phase 3 cluster randomized controlled trial.

We found that the participants in the intervention and control groups were
relatively similar demographically. However, participants at the control site
reported a higher household income and dietary intake of vegetables and various protein sources. Importantly, participants in the control group were more
likely to have an HIV-1 viral load below the limit of detection than participants
enrolled in the intervention group. Self-reported ART non-adherence was
uncommon in both groups. The lower proportion of non-detectable plasma viral
loads in the intervention group may be related to the fact that the intervention

group was economically worse off at baseline, and other studies have shown that markers of lower socioeconomic status are associated with delayed entry into care and worse HIV outcomes. In prior clinical programs in sub-Saharan Africa, viral load suppression ranges from 18% - 41% in observational studies (Hassan et al. 2014; Liegeois et al. 2012; Anude et al. 2013). Thus, our findings are relatively consistent with these results. Of note, this pilot study was not powered or designed to detect differences in viral load suppression by study arm. Rather the future trial will address variability and heterogeneity by randomizing approximately eight matched pair cluster (16 sites). We plan to match sites on a variety of criteria including sociodemographic factors, average rainfall, soil type, access to predominant source of surface water for irrigation (lake, vs. river, vs. shallow wells) and facility type (dispensary, vs. health center, vs. hospital).

Process evaluations garner in-depth knowledge around implementing complex interventions, which aids both interpretation of trial findings and informs potential replication (Oakley et al. 2006). Our process evaluation suggested robust acceptability of the *Shamba Maisha* intervention and trial, which aligns with the nearly 100% follow-up rate for both intervention and control participants at trial close. We also found that agricultural and financial training has strong advantages on their own, and that training should be integral to successful microfinance organizations. As anticipated, our qualitative data suggest that the microirrigation intervention can be laborsaving.

We also had significant challenges, most notably with regard to the microfinance component of the intervention. The partnering microfinance institution was unreliable, insolvent, and had a local branch closure during the trial. This is perhaps not surprising, given that past attempts to offer microfinance alongside HIV-related activities have experienced major challenges (Epstein 2006; Gregson et al. 2007). Researchers from the IMAGE intervention concluded that a strong microfinance partnership is indeed a pre-requisite to sustainably scaling this type of multisectoral intervention (Hargreaves et al. 2011). Thus, the replication of this trial in a larger setting should entail partnering with a microfinance institution that is financially stable and whose business model can accommodate Shamba Maisha. Participant concerns around loan repayment are an important ethical consideration; particularly in light of recent findings that microfinance may fail to provide enough income for financially disadvantaged households (Stewart et al. 2010; Barnes et al. 2001; Kaboski & Townsend 2008). As a result, we plan to consult local and global experts in microfinance in order to ensure that this component meets the needs of the study, and is designed with the potential for scale-up and sustainability is similar settings in East Africa.

These findings should be considered in light of design limitations. The pilot cluster-randomized design included two sites, and thus will need to be replicated in a larger cluster randomized trial. Second, qualitative process evaluations may suffer from respondent bias, particularly if participants held the impression that researchers were part of the intervention delivery team. We attempted to address this shortcoming by hiring separate in-depth interviewers that were not part of the primary study team. Lastly, the challenges with our microfinance organization made it difficult to assess the full potential of that component of the intervention.

We expect the pilot study will provide critical new evidence regarding the implementation of a multisectoral agricultural intervention aimed to improve food insecurity, household economic indicators and health among HIV-infected persons in rural Kenya. Following evaluation of the quantitative and qualitative results from this study, we plan to scale up this pilot to a cluster randomized trial to definitively assess the effectiveness of this intervention on the health of PLHIV using ART.

REFERENCES

1. About KickStart [http://www.kickstart.org/about-us/]
2. Agricultural initiatives for health in Haiti, Rwanda, and Lesotho
3. Anema A, Weiser SD, Fernandes KA, Brandson EK, Montaner JS, Hogg RS (2011) High prevalence of food insecurity among HIV-infected individuals receiving HAART in a resource-rich setting. AIDS Care 23(2):221–230
4. Antelman G, Smith Fawzi MC, Kaaya S, Mbwambo J, Msamanga GI, Hunter DJ, Fawzi WW (2001) Predictors of HIV-1 serostatus disclosure: a prospective study among HIV-infected pregnant women in Dar es Salaam. Tanzania AIDS 15(14):1865–1874
5. Anude CJ, Eze E, Onyegbutulem HC, Charurat M, Etiebet MA, Ajayi S, Dakum P, Akinwande O, Beyrer C, Abimiku A, Blattner W (2013) Immuno-virologic outcomes and immuno-virologic discordance among adults alive and on anti-retroviral therapy at 12 months in Nigeria. BMC Infect Dis 13:113
6. Ashburn K, Kerrigan D, Sweat M (2008) Micro-credit, women's groups, control of own money: HIV-Related negotiation among partnered Dominican women. AIDS behav 12(3):396–403
7. Babameto G, Kotler DP (1997) Malnutrition in HIV infection. Gastroenterol Clin North Am 26(2):393–415
8. Bangsberg DR, Hecht FM, Charlebois ED, Zolopa AR, Holodniy M, Sheiner L, Bamberger JD, Chesney MA, Moss A (2000) Adherence to protease inhibitors, HIV-1 viral load, and development of drug resistance in an indigent population. AIDS 14(4):357–366
9. Bangsberg DR, Hecht FM, Charlebois ED, Chesney MA, Moss A (2001a) Comparing objective measures of adherence to HIV antiretroviral therapy: Electronic medication monitors and unannounced pill counts. AIDS Behav 5:275–281

10. Bangsberg DR, Hecht FM, Clague H, Charlebois ED, Ciccarone D, Chesney M, Moss A (2001b) Provider assessment of adherence to HIV antiretroviral therapy. J Acquir Immune Defic Syndr 26(5):435–442

11. Bangsberg DR, Perry S, Charlebois ED, Clark RA, Roberston M, Zolopa AR, Moss A (2001c) Non-adherence to highly active antiretroviral therapy predicts progression to AIDS. AIDS 15(9):1181–1183

12. Bardsley-Elliot A, Plosker GL (2000) Nelfinavir: an update on its use in HIV infection. Drugs 59(3):581–620

13. Barnes C, Keogh E, Nemarundwe N (2001) Microfinance program clients and impact: An assessment of Zambuko Trust, Zimbabwe. AIMS, Washington, DC

14. Blumberg SJ, Dickey WC (2003) Prevalence of HIV risk behaviors, risk perceptions, and testing among US adults with mental disorders. J Acquir Immune Defic Syndr 32(1):77–79

15. Bolton P, Wilk CM, Ndogoni L (2004) Assessment of depression prevalence in rural Uganda using symptom and function criteria. Soc Psychiatry Psychiatr Epidemiol 39(6):442–447

16. Brandsma T (2003) Waterworks: Kenyan Farmers Are Getting A Big Boost From A Simple Piece Of Equipment. In: Newsweek. [http://www.newsweek.com/waterworks-137625]

17. Broadhead WE, Gehlbach SH, de Gruy FV, Kaplan BH (1988) The Duke-UNC Functional Social Support Questionnaire Measurement of social support in family medicine patients. Med Care 26(7):709–723

18. Byron E, Gillespie S, Nangami M (2008) Integrating nutrition security with treatment of people living with HIV: lessons from Kenya. Food Nutr Bull 29(2):87–97

19. Campbell AA, de Pee S, Sun K, Kraemer K, Thorne-Lyman A, Moench-Pfanner R, Sari M, Akhter N, Bloem MW, Semba RD (2009) Relationship of household food insecurity to neonatal, infant, and under-five child mortality among families in rural Indonesia. Food Nutr Bull 30(2):112–119

20. Cantrell RA, Sinkala M, Megazinni K, Lawson-Marriott S, Washington S, Chi BH, Tambatamba-Chapula B, Levy J, Stringer EM, Mulenga L, Stringer JSA (2008) A pilot study of food supplementation to improve adherence to antiretroviral therapy among food-insecure adults in Lusaka Zambia. J Acquir Immune Defic Syndr 49(2):190–195

21. Chatterton ML, Scott-Lennox J, Wu AW, Scott J (1999) Quality of life and treatment satisfaction after the addition of lamivudine or lamivudine plus loviride to zidovudine-containing regimens in treatment-experienced patients with HIV infection. Pharmacoeconomics 15(Suppl 1):67–74

22. Coates J, Swindale A, Bilinsky P (2006a) Household Food Insecurity Access Scale (HFIAS) for Measurement of Food Access: Indicator Guide. Food and Nutrition Technical Assistance. Academy for Educational Development, Washington, D.C, pp 1–30

23. Coates J, Frongillo EA, Rogers BL, Webb P, Wilde PE, Houser R (2006b) Commonalities in the experience of household food insecurity across cultures: what are measures missing? J Nutr 136(5):1438S–1448S

24. Collins C, Duffield A, Myatt M (2000) Adults Assessment of Nutritional Status in Emergency-Affected Populations In. United Nations, Administrative Committee on Coordination, Sub-committee on Nutrition, Geneva, Switzerland, pp 1–28

25. Dedoose Version 5.0.11, web application for managing, analyzing, and presenting qualitative and mixed method research data (2014). Los Angeles, CA: SocioCultural Research Consultants LLC.

26. Derogatis LR, Lipman RS, Rickels K, Uhlenhuth EH, Covi L (1974) The Hopkins Symptom Checklist (HSCL) A measure of primary symptom dimensions. Mod Probl Pharmacopsychiatry 7:79–110

27. Diagne A (1998) Impact of Access to Credit on Income and Food Security in Malawi. CND Discussion Paper No 46. International Food Policy Research Institute, Washington, D.C

28. Doocy S, Teferra S, Norell D, Burnham G (2005) Credit program outcomes: coping capacity and nutritional status in the food insecure context of Ethiopia. Soc Sci Med 60(10):2371–2382

29. Dunkle KL, Jewkes RK, Brown HC, Gray GE, McIntryre JA, Harlow SD (2004) Gender-based violence, relationship power, and risk of HIV infection in women attending antenatal clinics in South Africa. Lancet 363(9419):1415–1421

30. Dupas P, Robinson J (2010) Coping with political instability: micro evidence from Kenya's 2007 election crisis. Am Econ Rev 100(2):120–124

31. Dworkin SL, Blankenship K (2009) Microfinance and HIV/AIDS prevention: assessing its promise and limitations. AIDS Behav 13(3):462–469

32. Epstein H (2006) The underground economy of AIDS. Virginia Quart Rev 82(1):53–63

33. Evans DL, Ten Have TR, Douglas SD, Gettes DR, Morrison M, Chiappini MS, Brinker-Spence P, Job C, Mercer DE, Wang YL, Cruess D, Dube B, Dalen EA, Brown T, Bauer R, Petitto JM (2002) Association of depression with viral load, CD8 T lymphocytes, and natural killer cells in women with HIV infection. Am J Psychiatry 159(10):1752–1759

34. Food and Agriculture Organization (2010a) The State of Food Insecurity in the World—Addressing Food Insecurity in Protracted Crises. Food and Agriculture Organization of the United Nations, Rome

35. Ferro-Luzzi A, Sette S, Franklin M, James WP (1992) A simplified approach of assessing adult chronic energy deficiency. Eur J Clin Nutr 46(3):173–186

36. Fields-Gardner C, Fergusson P (2004) Position of the American Dietetic Association and Dietitians of Canada: nutrition intervention in the care of persons with human immunodeficiency virus infection. J Am Diet Assoc 104(9):1425–1441

37. Food and Agriculture Organization (2010b) The State of Food Insecurity in the World—Addressing Food Insecurity in Protracted Crises. Food and Agriculture Organization of the United Nations, Rome

38. Food and Nutrition Technical Assistance. HIV/AIDS: A Guide For Nutritional Care and Support [http://www.fantaproject.org/publications/HIVguide.shtml]

39. Frongillo EA, Nanama S (2006) Development and validation of an experience-based measure of household food insecurity within and across seasons in northern Burkina Faso. J Nutr 136(5):1409S–1419S

40. Gelberg L, Gallagher TC, Andersen RM, Koegel P (1997) Competing priorities as a barrier to medical care among homeless adults in Los Angeles. Am J Public Health 87(2):217–220

41. Gelberg L, Andersen RM, Leake BD (2000) The Behavioral Model for Vulnerable Populations: application to medical care use and outcomes for homeless people. Health Serv Res 34(6):1273–1302

42. Global report (2013) UNAIDS report on the global AIDS epidemic. Joint United Nations Programme on HIV/AIDS, Geneva Switzerland

43. Gomez B, Lister A, Wiese M (2004) Micro-credit: giving life through work to PLWHA. International AIDS Conference: 2004, Bangkok, Thailand

44. Goudge J, Ngoma B (2011) Exploring antiretroviral treatment adherence in an urban setting in South Africa. J Public Health Policy 32(Suppl 1):S52–S64
45. Government of Kenya (2008) District Development Plan: Migori. District Commissioner's Office, Government of Kenya, Migori, Kenya
46. Gregson S, Adamson S, Papaya S, Mundondo J, Nyamukapa CA, Mason PR, Garnett GP, Chandiwana SK, Foster G, Anderson RM (2007) Impact and Process Evaluation of Integrated Community and Clinic-Based HIV-1 Control: A Cluster-Randomised Trial in Eastern Zimbabwe. PLoS Medicine 4(3):e102
47. Grosh M, Glewwe P (1998) Data Watch: the World Bank's Living Standard Measurement Study household surveys. J Econ Perspect 12(1):187–196
48. Gustavson L, Lam W, Bertz R, Hsu A, Rynkiewicz K, Ji Q, Ghosh S, Facey I, Bernstein B, Sun E (2000) Assessment of the bioequivalence and food effects for liquid and soft gelatin capsule co-formulations of ABT-378/ritonavir (ABT-378/r) in healthy subjects. 40th Interscience Conference on Antimicrobial Agents and Chemotherapy: 2000, Toronto, Canada
49. Hargreaves J, Hatcher A, Strange V, Phetla G, Busza J, Kim J, Watts C, Morison L, Porter J, Pronyk P, Bonnell C (2010) Process evaluation of the Intervention with Microfinance for AIDS and Gender Equity (IMAGE) in rural South Africa. Health Educ Res 25(1):27–40
50. Hargreaves J, Hatcher A, Busza J, Strange V, Phetla G, Kim J, Watts C, Porter JD, Pronyk P, Bonell C (2011) What happens after a trial? Replicating a cross-sectoral intervention addressing the social determinants of health: the case of the Intervention with Microfinance for AIDS and Gender Equity (IMAGE) in South Africa. In: Blas E, Sommerfeld J, Kurup AS (eds) Social determinants approaches to public health. World Health Organization, Geneva, pp 147–159
51. Hassan AS, Nabwera HM, Mwaringa SM, Obonyo CA, Sanders EJ, de Wit Rinke TF, Cane PA, Berkley JA (2014) HIV-1 virologic failure and acquired drug resistance among first-line antiretroviral experienced adults at a rural HIV clinic in coastal Kenya: a cross-sectional study. AIDS Res Ther 11(1):9
52. Hatcher AM, Tsai AC, Kumbakumba E, Dworkin SL, Hunt PW, Martin JN, Clark G, Bangsberg DR, Weiser SD (2012) Sexual relationship power and depression among HIV-infected women in Rural Uganda. PLoS One 7(12):e49821
53. Hiarlaithe MO, Grede N, de Pee S, Bloem M (2014) Economic and Social Factors are Some of the Most Common Barriers Preventing Women from Accessing Maternal and Newborn Child Health (MNCH) and Prevention of Mother-to-Child Transmission (PMTCT) Services: A Literature Review. AIDS Behav 18 Suppl 5:S516–S530
54. Hoddinott J, Yohannes Y (2002) Dietary diversity as a household food security indicator. Food and Nutrition Technical Assistance Project, Academy for Educational Development, Washington, D.C
55. Ickovics JR, Hamburger ME, Vlahov D, Schoenbaum EE, Schuman P, Boland RJ, Moore J (2001) Mortality, CD4 cell count decline, and depressive symptoms among HIV-seropositive women: longitudinal analysis from the HIV Epidemiology Research Study. JAMA 285(11):1466–1474
56. James WP, Ferro-Luzzi A, Waterlow JC (1988) Definition of chronic energy deficiency in adults Report of a working party of the International Dietary Energy Consultative Group. Eur J Clin Nutr 42(12):969–981

57. Jewkes RK, Dunkle K, Nduna M, Shai N (2010) Intimate partner violence, relationship power inequity, and incidence of HIV infection in young women in South Africa: a cohort study. Lancet 376(9734):41–48

58. Johannessen A, Naman E, Ngowi BJ, Sandvik L, Matee MI, Aglen HE, Gundersen SG, Bruun JN (2008) Predictors of mortality in HIV-infected patients starting antiretroviral therapy in a rural hospital in Tanzania. BMC Infect Dis 8:52

59. Kaboski JP, Townsend RM (2008) A structural evaluation of a large-scale quasi-experimental microfinance initiative. In: MIT Department of Economics Working Paper No. Mass Institute of Technology, Boston, pp 09–12

60. Kalichman SC, Simbayi LC, Cloete A, Mthembu PP, Mkhonta RN, Ginindza T (2009) Measuring AIDS stigmas in people living with HIV/AIDS: the Internalized AIDS-Related Stigma Scale. AIDS Care 21(1):87–93

61. Kalichman SC, Cherry C, Amaral C, White D, Kalichman MO, Pope H, Swetsze C, Jones M, Macy R (2010a) Health and treatment implications of food insufficiency among people living with HIV/AIDS, Atlanta. Georgia J Urban Health 87(4):631–641

62. Kalichman SC, Cherry C, Amaral CM, Swetzes C, Eaton L, Macy R, Grebler T, Kalichman MO (2010b) Adherence to antiretroviral therapy and HIV transmission risks: implications for test-and-treat approaches to HIV prevention. AIDS Patient Care STDS 24(5):271–277

63. Kenya National Bureau of Statistics and ICF Macro (2010) Kenya Demographic and Health Survey 2008–09. Kenya National Bureau of Statistics and ICF Macro, Calverton, Maryland

64. Kihia JK, Kamau RN (1999) Super-Money Maker pressure pedal pump impact assessment in utilisation, job creation and income generation. In. ApproTEC, Inc., Nairobi

65. Kim J, Pronyk P, Barnett T, Watts C (2008) Exploring the role of economic empowerment in HIV prevention. Aids 22:S57–S71

66. Kotler DP, Tierney AR, Brenner SK, Couture S, Wang J, Pierson RN Jr (1990) Preservation of short-term energy balance in clinically stable patients with AIDS. Am J Clin Nutr 51(1):7–13

67. Larson B, Fox M, Rosen S, Bii M, Sigei C, Shaffer D, Sawe F, Wasunna M, Simon J (2008) Early effects of antiretroviral therapy on work performance: preliminary results from a cohort study of Kenyan agricultural workers. AIDS 22(3):421

68. Lee JS, Frongillo EA Jr (2001) Nutritional and health consequences are associated with food insecurity among U.S. elderly persons. J Nutr 131(5):1503–1509

69. Lewis Kulzer J, Penner JA, Marima R, Oyaro P, Oyanga AO, Shade SB, Blat CC, Nyabiage L, Mwachari CW, Muttai HC, Bukusi EA, Cohen CR (2012) Family model of HIV care and treatment: a retrospective study in Kenya. J Int AIDS Soc 15(1):8

70. Liegeois F, Vella C, Eymard-Duvernay S, Sica J, Makosso L, Mouinga-Ondeme A, Mongo AD, Boue V, Butel C, Peeters M, Gonzalez JP, Delaporte E, Rouet F (2012) Virological failure rates and HIV-1 drug resistance patterns in patients on first-line antiretroviral treatment in semirural and rural Gabon. J Int AIDS Soc 15(2):17985

71. Macallan DC, Noble C, Baldwin C, Jebb SA, Prentice AM, Coward WA, Sawyer MB, McManus TJ, Griffin GE (1995) Energy expenditure and wasting in human immunodeficiency virus infection. N Engl J Med 333(2):83–88

72. Mahlungulu S, Grobler LA, Visser ME, Volmink J (2007) Nutritional interventions for reducing morbidity and mortality in people with HIV. Cochrane Database Syst Rev 3, CD004536

73. Mamlin J, Kimaiyo S, Lewis S, Tadayo H, Jerop FK, Gichunge C, Petersen T, Yih Y, Braitstein P, Einterz R (2009) Integrating nutrition support for food-insecure patients and their dependents into an HIV care and treatment program in Western Kenya. Am J Public Health 99(2):215–221

74. Mast TC, Kigozi G, Wabwire-Mangen F, Black R, Sewankambo N, Serwadda D, Gray R, Wawer M, Wu AW (2004) Measuring quality of life among HIV-infected women using a culturally adapted questionnaire in Rakai district Uganda. AIDS Care 16(1):81–94

75. Mbugua S, Andersen N, Tuitoek P, Yeudall F, Sellen D, Karanja N, Cole D, Njenga M, Prain G (2008) Assessment of food security and nutrition status among households affected by HIV/AIDS in Nakuru Municipality, Kenya. XVII International AIDS Conference, Mexico City

76. McIntyre D, Thiede M, Dahlgren G, Whitehead M (2006) What are the economic consequences for households of illness and of paying for health care in low-and middle-income country contexts? Soc Sci Med 62(4):858–865

77. McMahon JH, Wanke CA, Elliott JH, Skinner S, Tang AM (2011) Repeated assessments of food security predict CD4 change in the setting of antiretroviral therapy. J Acquir Immune Defic Syndr 58(1):60–63

78. Miller CL, Bangsberg DR, Tuller DM, Senkungu J, Kawuma A, Frongillo EA, Weiser SD (2011) Food insecurity and sexual risk in an HIV endemic community in Uganda. AIDS Behav 15(7):1512–1519

79. Nagata JM, Magerenge RO, Young SL, Oguta JO, Weiser SD, Cohen CR (2012) Social determinants, lived experiences, and consequences of household food insecurity among persons living with HIV/AIDS on the shore of Lake Victoria, Kenya. AIDS Care 24(6):728–736

80. Njenga M, Karanja N, Gathuru K, Mbugua S, Fedha N, Ngoda B (2009) The role of women-led micro-farming activities in combating HIV/AIDS in Nakuru, Kenya. In: Hovorka A, de Zeeuw H, Njenga M (eds) Women feeding cities: mainstreaming gender in urban agriculture and food security. Practical Action Publishing, Warwickshire, Rugby, UK

81. Normen L, Chan K, Braitstein P, Anema A, Bondy G, Montaner JS, Hogg RS (2005) Food insecurity and hunger are prevalent among HIV-positive individuals in British Columbia. Canada J Nutr 135(4):820–825

82. Nutrition and HIV/AIDS (2001) Statement by the Administrative Committee on Coordination, Sub-Committee on Nutrition at its 28th Session. United Nations Administrative Committee on Coordination, Sub-Committee on Nutrition, Nairobi, Kenya

83. Oakley A, Strange V, Bonell C, Allen E, Stephenson J (2006) Process evaluation in randomised controlled trials of complex interventions. British Medical Journal 332(7538):413–416

84. Ochai R (2008) HIV, livelihoods, nutrition and health research. Global Ministerial Forum on Research for Health: 2008, Bamako, Mali

85. Oyugi J, Byakika-Tusiime J, Charlebois E, Kityo C, Mugerwa R, Mugyenyi P, Bangsberg D (2004) Multiple validated measures of adherence indicate high levels of adherence to generic HIV antiretroviral therapy in a resource-limited setting. J Acquir Immune Defic Syndr 36(5):1100–1102

86. Oyugi JH, Byakika-Tusiime J, Ragland K, Laeyendecker O, Mugerwa R, Kityo C, Mugyenyi P, Quinn TC, Bangsberg DR (2007) Treatment interruptions predict resistance in HIV-positive individuals purchasing fixed-dose combination antiretroviral therapy in Kampala. Uganda Aids 21(8):965–971

87. Pandit JA, Sirotin N, Tittle R, Onjolo E, Bukusi EA, Cohen CR (2010) Shamba Maisha: a pilot study assessing impacts of a micro-irrigation intervention on the health and economic wellbeing of HIV patients. BMC Pub Health 10:245

88. Parienti JJ, Massari V, Descamps D, Vabret A, Bouvet E, Larouze B, Verdon R (2004) Predictors of virologic failure and resistance in HIV-infected patients treated with nevirapine- or efavirenz-based antiretroviral therapy. Clin Infect Dis 38(9):1311–1316

89. Parienti JJ, Das-Douglas M, Massari V, Guzman D, Deeks SG, Verdon R, Bangsberg DR (2008) Not all missed doses are the same: sustained NNRTI treatment interruptions predict HIV rebound at low-to-moderate adherence levels. PLoS ONE 3(7):e2783

90. Petersen ML, Sinisi SE, van der Laan MJ (2006) Estimation of direct causal effects. Epidemiology 17(3):276–284

91. Physical Status (1995) The Use and Interpretation of Anthropometry. Report of a WHO Expert Committee. WHO Technical Report Series 854. In. World Health Organization, Geneva, Switzerland

92. Pronyk PM, Hargreaves JR, Kim JC, Morison LA, Phetla G, Watts C, Busza J, Porter JD (2006) Effect of a structural intervention for the prevention of intimate-partner violence and HIV in rural South Africa: a cluster randomised trial. Lancet 368(9551):1973–1983

93. Pulerwitz J, Gortmaker SL, DeJong W (2000) Measuring sexual relationship power in HIV/STD research. Sex Roles 42(7–8):637–660

94. Robinson J, Yeh E (2011) Transactional sex as a response to risk in western Kenya. Am Econ J Appl Econ 3:35–64

95. Rose D, Oliveira V (1997) Nutrient intakes of individuals from food-insufficient households in the United States. Am J Public Health 87(12):1956–1961

96. Russell S (2004) The economic burden of illness for households in developing countries: a review of studies focusing on malaria, tuberculosis, and human immunodeficiency virus/acquired immunodeficiency syndrome. Am J Trop Med Hygiene 71(2 Suppl):147–155

97. Schuler SR, Hashemi SM (1994) Credit Programs, Womens Empowerment, And Contraceptive Use In Rural Bangladesh. Stud Family Plan 25(2):65–76

98. Siedner MJ, Tsai AC, Dworkin S, Mukiibi NF, Emenyonu NI, Hunt PW, Haberer JE, Martin JN, Bangsberg DR, Weiser SD (2012) Sexual relationship power and malnutrition among HIV-positive women in rural Uganda. AIDS Behav 16(6):1542–1548

99. Stack JA, Bell SJ, Burke PA, Forse RA (1996) High-energy, high-protein, oral, liquid, nutrition supplementation in patients with HIV infection: effect on weight status in relation to incidence of secondary infection. J Am Diet Assoc 96(4):337–341

100. Stevens JE (2002) Martin makes a middle class: Stanford grad Martin Fisher has gone low-tech in search of solutions for Kenyan farmers. SF Chronicle Magazine, San Francisco, p 18

101. Stewart R, Rooyen C, Dickson K, Majoro M, Wet T (2010) What is the impact of microfinance on poor people?: a systematic review of evidence from sub-Saharan Africa. In. EPPI-Centre, Social Science Research Unit, Institute of Education, University of London, London

102. Stringer JS, Zulu I, Levy J, Stringer EM, Mwango A, Chi BH, Mtonga V, Reid S, Cantrell RA, Bulterys M, Saag MS, Marlink RG, Mwinga A, Ellerbrock TV, Sinkala M (2006) Rapid scale-up of antiretroviral therapy at primary care sites in Zambia: feasibility and early outcomes. JAMA 296(7):782–793

103. Swindale A, Bilinsky P (2006) Development of a universally applicable household food insecurity measurement tool: process, current status, and outstanding issues. J Nutr 136(5):1449S–1452S

104. Sztam KA, Fawzi WW, Duggan C (2010) Macronutrient supplementation and food prices in HIV treatment. J Nutr 140(1):213S–223S

105. Tsai AC, Bangsberg DR, Emenyonu N, Senkungu JK, Martin JN, Weiser SD (2011) The social context of food insecurity among persons living with HIV/AIDS in rural Uganda. Soc Sci Med 73(12):1717–1724

106. Tsai AC, Bangsberg DR, Frongillo EA, Hunt PW, Martin JN, Weiser SD (2012) Food insecurity, depression and the modifying role of social support among people living with HIV/AIDS in Rural Uganda. Soc Sci Med 74(12):2012–2019

107. Tsai AC, Bangsberg DR, Kegeles SM, Katz IT, Muzoora C, Martin JN, Weiser SD (2013) Internalized stigma, disease progression, and serostatus disclosure among people living with HIV/AIDS in rural Uganda. Ann Behav Med 46(3):285–294

108. Tucker JS, Burnam MA, Sherbourne CD, Kung FY, Gifford AL (2003) Substance use and mental health correlates of nonadherence to antiretroviral medications in a sample of patients with human immunodeficiency virus infection. Am J Med 114(7):573–580

109. Tuller DM, Bangsberg DR, Senkungu J, Ware NC, Emenyonu N, Weiser SD (2010) Transportation costs impede sustained adherence and access to HAART in a clinic population in southwestern Uganda: a qualitative study. AIDS Behav 14(4):778–784

110. UNAIDS Policy Brief (2008) HIV/AIDS, HIV, Food Security and Nutrition.

111. Wang EA, McGinnis KA, Fiellin DA, Goulet JL, Bryant K, Gibert CL, Leaf DA, Mattocks K, Sullivan LE, Vogenthaler N, Sullivan LE, Vogenthaler N, Justice AC (2011) Food insecurity is associated with poor virologic response among HIV-infected patients receiving antiretroviral medications. J Gen Intern Med 26(9):1012–1018

112. Webb-Girard A, Cherobon A, Mbugua S, Kamau-Mbuthia E, Amin A, Sellen DW (2012) Food insecurity is associated with attitudes towards exclusive breastfeeding among women in urban Kenya. Matern Child Nutr 8(2):199–214

113. Weinhardt LS, Galvao LW, Stevens PE, Masanjala WH, Bryant C, Ng'ombe T (2009) Broadening research on microfinance and related strategies for HIV prevention: commentary on Dworkin and Blankenship (2009). AIDS Behav 13(3):470–473

114. Weiser SD, Wolfe WR, Bangsberg DR (2004) The HIV epidemic among individuals with mental illness in the United States. Curr HIV/AIDS Rep 1(4):186–192

115. Weiser SD, Leiter K, Bangsberg DR, Butler LM, Percy-de Korte F, Hlanze Z, Phaladze N, Iacopino V, Heisler M (2007) Food insufficiency is associated with high-risk sexual behavior among women in Botswana and Swaziland. PLoS Med 4(10):1589–1597, discussion 1598

116. Weiser SD, Frongillo EA, Ragland K, Hogg RS, Riley ED, Bangsberg DR (2009a) Food insecurity is associated with incomplete HIV RNA suppression among homeless and marginally housed HIV-infected individuals in San Francisco. J Gen Intern Med 24(1):14–20

117. Weiser SD, Fernandes KA, Brandson EK, Lima VD, Anema A, Bangsberg DR, Montaner JS, Hogg RS (2009b) The association between food insecurity and mortality among HIV-infected individuals on HAART. J Acq Immune Def Syndromes 52(3):342–349

118. Weiser SD, Bangsberg DR, Kegeles S, Ragland K, Kushel MB, Frongillo EA (2009c) Food insecurity among homeless and marginally housed individuals living with HIV/AIDS in San Francisco. AIDS Behav 13(5):841–848

119. Weiser S, Fernandes K, Anema A, Brandson E, Lima V, Montaner J, Hogg R (2009d) Food insecurity as a barrier to antiretroviral therapy (ART) adherence among HIV-infected individuals in British Columbia (2009d). 5th IAS Conference on HIV Pathogenesis, Treatment and Prevention, Cape Town, South Africa

120. Weiser SD, Tsai AC, Senkungu J, Emenyonu N, Kawuma A, Hunt P, Martin J, Bangsberg DR (2010a) The impact of low sexual relationship power on viral load suppression among women receiving antiretroviral therapy in Mbarara, Uganda. 17th Conference on Retroviruses and Opportunistic Infections, San Francisco
121. Weiser SD, Tuller DM, Frongillo EA, Senkungu J, Mukiibi N, Bangsberg DR (2010b) Food insecurity as a barrier to sustained antiretroviral therapy adherence in Uganda. PLoS One 5(4):e10340
122. Weiser SD, Young SL, Cohen CR, Kushel MB, Tsai AC, Tien PC, Hatcher AM, Frongillo EA, Bangsberg DR (2011) Conceptual framework for understanding the bidirectional links between food insecurity and HIV/AIDS. Am J Clin Nutr 94(6):1729S–1739S
123. Weiser SD, Tsai AC, Gupta R, Frongillo EA, Kawuma A, Senkungu J, Hunt PW, Emenyonu NI, Mattson JE, Martin JN, Bangsberg DR (2012) Food insecurity is associated with morbidity and patterns of healthcare utilization among HIV-infected individuals in a resource-poor setting. AIDS 26(1):67–75
124. Wolfe WR, Weiser SD, Leiter K, Steward WT, Percy-de Korte F, Phaladze N, Iacopino V, Heisler M (2008) The impact of universal access to antiretroviral therapy on HIV stigma in Botswana. Am J Public Health 98(10):1865–1871
125. World Bank (2010) Yes, Africa can: success stories from a dynamic continent. The World Bank, Washington, D.C
126. World Food Program (2003) Programming in the Era of AIDS: WFP's Response to HIV/AIDS, WFP/EB.1/2003/4-B.
127. Wu AW, Rubin HR, Mathews WC, Ware JE Jr, Brysk LT, Hardy WD, Bozzette SA, Spector SA, Richman DD (1991) A health status questionnaire using 30 items from the Medical Outcomes Study Preliminary validation in persons with early HIV infection. Med Care 29(8):786–798
128. Zachariah R, Fitzgerald M, Massaquoi M, Pasulani O, Arnould L, Makombe S, Harries AD (2006) Risk factors for high early mortality in patients on antiretroviral treatment in a rural district of Malawi. Aids 20(18):2355–2360

PART IV
Food Security and Obesity and Diabetes

Challenges of Diabetes Self-Management in Adults Affected by Food Insecurity in a Large Urban Centre of Ontario, Canada

Justine Chan, Margaret DeMelo, Jacqui Gingras, and Enza Gucciardi

9.1 INTRODUCTION

By 2020, the cost to treat diabetes mellitus, its complications, and its associated loss of productivity and life will exceed CAN$19 billion a year [1]. Compelling evidence supports the benefits of intensive glycemic, lipid, and blood pressure control in the prevention and management of diabetes complications [2]. However, diabetes self-management is a challenge for many people as it involves learning and adopting self-care and self-monitoring practices [3]. For instance, medical nutrition therapy, a cornerstone in diabetes management, is one of the most challenging aspects of management [4, 5], encompassing not only healthy eating, but also insulin dose adjustment to carbohydrate consumed, as well as prevention and treatment of hypoglycaemia.

Household food insecurity is significantly more common among Canadians with diabetes (9.3%) compared to Canadians without diabetes (6.8%) [6], which is similar to that of other countries [7, 8]. Food security exists "when all people, at all times, have physical and economic access to sufficient, safe, and nutritious food to meet their dietary needs and food preferences for an active and healthy life" [9]. Notably rooted in poverty, food insecurity for people with diabetes poses additional challenges, principally, the lack of adequate and appropriate food and the effect on diabetes management [10]. Studies of adults who are food insecure show that they are more likely to have poorer health, inferior social support, and more comorbid conditions (e.g., obesity, high blood pressure, heart disease, and allergies) [11] as well as greater psychological distress and unhealthy behaviours, including smoking, physical inactivity, and low consumption of fruits and vegetables [6, 11]. Specifically, food insecure adults with diabetes are more likely to have poor glycemic control, long-term complications, and severe and frequent hypoglycaemia [12–16]. While it is clear that those that are living with diabetes and are food insecure experience more health problems and poorer overall health, knowledge on how they cope and manage with this health intersection and information on their lived experiences to date have been primarily absent in the literature. This type of knowledge can provide insight on how to better support and care for this population.

Using in-depth interviews with adults living with diabetes and food insecurity, our research paper explores these lived experiences and tries to understand how food insecurity affects people's ability to manage their diabetes. The semistructured interview guide was based on the Social Determinants of Health Framework, given the significant influence of material deprivation and financial constraints on diabetes management decisions [17]. The findings will be novel as there is very limited qualitative research in this area.

9.2　MATERIALS AND METHODS

9.2.1　Qualitative Methodology

There is little contextual research on how and why individuals with diabetes coping with food insecurity are more likely to practice unhealthy behaviours and endure greater psychological distress. Using six phases of analysis, we chose qualitative, deductive thematic analysis, as it is flexible in its approach to extract themes from participants' accounts [18].

9.2.2   Participant Recruitment

With the help of diabetes educators and flyers promoting the study, we conducted one-on-one interviews with clients from the local community, three
community health centres, and a community-based diabetes education centre
serving a low-income population in the Greater Toronto Area, Ontario, Canada.
Eligibility criteria included a diagnosis of type 1 or 2 diabetes, English speaking,
and having experienced food insecurity in the past year. Three questions were
adapted from the Household Food Security Survey Module (HFSSM) [19] to
identify participants: (1) In the past year, were you ever not able to buy your
basic foods, such as fruits and vegetables? (2) In the past year, were you ever
not able to buy your favourite foods? (3) In the past year, did you ever have to
eat less than you felt you should have because of a limited budget? Eligibility
was based on at least one affirmative response to the above questions. Homeless
individuals were excluded, as the complexity of psychosocial vulnerabilities and
mental health issues would extend far beyond the scope of our research question. See Table 1 for our sample demographics. Participant recruitment ended
when theme saturation was achieved [20].

TABLE 9.1   Sociodemographic characteristics of participants.

	Number (n = 21)
Gender	
Women	10
Men	11
Ethnicity	
Caucasian	12
Caribbean	5
African	2
Middle Eastern	2
Place of birth	
Canada	12
Outside of Canada	9
Age	
20–30	1
31–40	1
41–50	7
51–60	9
61–70	3

TABLE 9.1   *(Continued)*

	Number (n = 21)
Marital status	
Married/common-law	2
Single	9
Divorced/separated/widowed	10
Education	
High school or less	7
University or less	4
College or less	9
Graduate school	1
Duration of diabetes	
≤5 years	11
6–10 years	4
11–20 years	2
21–30 years	2
>30 years	2
Type of diabetes	
Type 1	1
Type 2	20

Ethics approval for this research study was obtained from Ryerson University's Research Ethics Board in Ontario, Canada (REB 2008-294). Initial data collection for this study was completed in 2009-2010. The four-member research team are comprised of one academic expert in diabetes (Enza Gucciardi), one expert in qualitative research (Jacqui Gingras), and two practice experts in diabetes (Justine Chan, Margaret DeMelo).

9.2.3   Procedure for Data Analysis

Eligibility, demographic information, and verbal informed consent were confirmed by the lead author prior to the interview. In total, 21 individual face-to-face interviews were conducted, lasting 30 to 90 minutes.

The semistructured interview guide is as follows.

Introduction

- Can you remember when you were first diagnosed with diabetes? What was it like?
- How did you react to the news?

Health Services

Think back to when you first learned about what you would need to do to manage your diabetes. How was that experience for you?

- What did the doctor/nurse/and so forth tell you about what you had to do to manage your diabetes? What was it like learning about what you needed to do in terms of food, physical activity, and medication? What challenges did you face in trying to follow the recommendations of your physician or the diabetes health care team?
- Can you tell me when you first found it difficult to buy or find food? Was this before or after you found out you had diabetes?

Income and Social Status

How did you manage through the times when you found it hard to buy or find food? How has this affected you in the past year? If they have kids or living with family: When providing meals for other people in the household, what were your major concerns?

Social Environments/Social Support Networks

Were you able to get the help/assistance/understanding/support you felt you needed during these times?

- What support was available to you at this time (formal support from health/social care providers, family, friends, church group, etc.)? Would you tell a bit me about these experiences? What supports really made a difference for you; what was most helpful to you? What more can they (heath care providers, government, etc.) do to better help you?
- How have concerns around food affected your daily life?

Personal Health Practices and Coping Skills

Some of the things we've been talking about sound like they may have been very difficult/challenging for you. What have been the biggest challenges overall? What have been your "survival" strategies through all this? How confident are you now in your ability to manage your diabetes?

Cool-Down/Wrap-Up

Looking back on your own experiences, what would have made this whole journey easier for you?

- If you could send a message to other people in your situation, what would it be?
- What about healthcare providers? What would you like them to know?
- Is there anything else about this experience that was important to you that we haven't talked about?
- The responses you have provided may lead to some more questions. If so, can we contact you for a follow-up interview?

Probes were used to encourage participants to elaborate on their experiences. Participants received CAN$30 honorarium at the end of their interviews. All interviews were digitally recorded and transcribed verbatim. Following data collection, we removed all participant identifiers to protect anonymity.

Our analysis process required six phases [21], beginning with the research team reviewing interview transcripts, while making notes for potential codes to return to at a later stage in analysis. The second phase involved developing preliminary codes, 50 in total, to capture as many potential themes as possible. We grouped transcript excerpts under these codes both manually and by NVivo data management software. Following phase two, we created clusters of codes and grouped them into ten major themes using the repetition technique where recurring topics generated the most relevant ideas [21]. In the third phase, we used thematic networks [22] to systematically present the study findings by listing our themes from specific to broad, that is, basic, organizing, and global themes, respectively. In the fourth phase, we met frequently to further refine our themes and triangulate the data. The basic themes were then grouped under three overarching themes. In the fifth phase, we refined theme wording to capture the meaning of what was said. The sixth phase involved writing and revising the manuscript report, while the research team carefully considered the most meaningful extracts.

9.3 · RESULTS

Our data analysis produced three main themes that captured the experiences of people with diabetes who are food insecure: (1) barriers to preparing and accessing appropriate food, (2) social isolation, and (3) enhancing agency and resilience (see Figure 1).

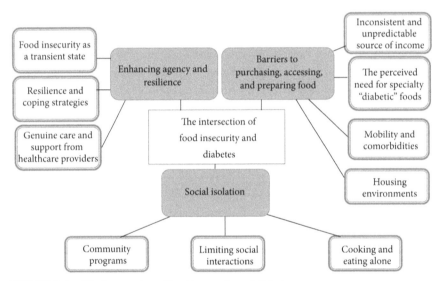

FIGURE 9.1   The intersection of food insecurity and diabetes.

9.3.1   Themes and Additional Supporting Quotes

Theme 1 (barriers to preparing and accessing appropriate food). Most participants could not afford foods appropriate for their diabetes management. Always buying "healthy foods," counting carbohydrates, or tracking serving amounts were unrealistic approaches to meal planning because food supplies were often erratic due to an inconsistent and unpredictable source of income. When grocery shopping, their goal was to buy whatever cost the least (see Interviewee 1). Participants also worried about not having enough food to eat and described the additional challenges brought on by a limited budget (see Interviewee 2). Meal planning was difficult and many depended on other resources, such as food banks and community kitchens. Inappropriate foods available at food banks (i.e., high in starch, salt, and sugar) were voiced (see Interviewees 4 and 5). Unfortunately, participants still struggled with access to these food sources (see Interviewee 3). For example, a single mother modified her work hours with the food bank's operating hours. Access to food was often better at the beginning of the month, resulting in more erratic blood sugars by month's end.

Housing environments presented barriers to food preparation. Many did not own a stove, resorting to microwaveable foods (see Interviewee 6). A lack of proper cooking facilities resulted in greater use of higher sodium foods such as processed and canned foods.

Misperception of the type of foods recommended for diabetes management was widely common. Participants discussed how specialty foods, such as those containing artificial sweeteners or labeled "diabetic," were perceived as "better," expensive, and, contrary to current nutrition recommendation, understood to be necessary for the management of diabetes (see Interviewees 7 and 9).

The barriers resulting from the intersection of diabetes and food insecurity were especially evident for those who had to cope with debilitating comorbidities. Retinopathy and neuropathy, in particular, greatly reduced their access and selection of both fresh and appropriate foods (see Interviewee 8). For example, many voiced the challenge of no longer being able to travel to food banks or grocery stores and had difficulty selecting and preparing food (see Interviewee 2).

Interviewee 1: I don't have a lot of money...so I'll buy junk food, instead of real food...because the junk food is cheaper.

Interviewee 2: I'm on [disability benefits] at the moment, so sometimes I find it's a bit hard...I have to be very conscious of...and trying to think of not just buying food but planning meals and planning food throughout the month that will last throughout the month...sometimes I worry about running out of food and, because I'm diabetic, I can't just skip meals...I find I have to be more conscious of what I eat and when I eat...I find that a bit troubling.

Interviewee 3: It's all controlled by how much money you make...how much you pay rent. Most places [food banks] say you can come once every two weeks – and they give you one shopping bag of groceries and that's supposed to last you two weeks. If you don't have the funds, how are you supposed to eat? So you have to go to the food kitchens...you have your days kept busy running back and forth to find places to eat... some days you might not make it there... so you don't eat properly.

Interviewee 3: A lot of my food comes from the food bank. They only give you certain types of foods that don't really help you with diabetes, more or less go against your diabetes—a lot of sugar, cookies, stuff like that.

Interviewee 4: It's mostly pastas, heavy in starch, pasta, rice, pasta sauce... and cereals, stuff that's got sugar in it. There's not an alternative they give you.

Interviewee 5: You don't have enough fruits – juice is no good. The fruits are better, but you have zero access. They have fruit roll ups, they have junk food which I don't eat, so I give back most of the stuff.

Interviewee 6: The fact that I have the necessary facilities to cook... I've got a stove... but I never use it because... the person before [was]... I found this little thing and it'd blow up on me... I mean it looks nice, it's kept clean and it's spotless inside, but so is a hand grenade before you pull the pin too... I bought my own microwave... I use everything microwave or I eat out of a can cold.

Interviewee 7: I have to get... artificial sweetener now, too, and that's expensive compared to other stuff... I get the diet soft drinks, and I get juices. But then there's a lot of sugar in the juices, so I try and mix it with water, and then I find that not a lot of grocery stores have just a specific section. I've noticed that some of them, like the drug stores... have a certain section. Some of the items that are in the diabetic aisles... they are expensive too, right?

Interviewee 9: It'd be easier for me if I could just buy regular food, I don't have to buy special stuff that's low sugar and stuff. If I want it, I'd have to pay more for it.

Interviewee 8: For the past 4, 5 years... all the eye problems and everything else... it made it difficult to do things. You can't manage. You can't see to cook... the finger gets numb because of the diabetes. The nerves and the fingers, you cannot hold things... and even when you're buying stuff... you're holding it, but you can't really tell whether it's... good or [if] it's not good... I just... use the canned goods.

Interviewee 2: I hate going [to] the grocery store and getting... loads and loads... of groceries. I'd much prefer to shop every two or three days, but that's not really practical in terms of getting enough food and... making sure it keeps fresh…it's hard when…I've got some physical problems and hauling it around…is…difficult.

Theme 2 (social isolation). Most participants were single and described the impact of food and eating alone on their well-being. Socializing incurred more cost on travel and clothes (see Interviewee 5). Because lack of funds limited social interactions, many felt isolated and depressed (see Interviewee 6).

While some considered eating out a special treat, especially for those who felt lonely and isolated, the impact of limited finances and having diabetes further restricted social interactions. Feeling the need to justify their food and beverage choices and to eat and drink differently from their friends were apparent barriers (see Interviewee 9).

Community food initiatives such as community kitchens and gardens, volunteering at food distribution centres, connecting to a church group, and attending drop-in meal programs were all means of connecting with others and were considered a valuable resource that helped to ameliorate both food insecurity and social isolation (see Interviewees 10 and 11).

Interviewee 5: I've…been inverting myself more to a cocoon?…I've stopped socializing…don't feel like it…cost of going, even if it's just subway or gas, it's money. So…I finally went to [an event] last week, because…I really had the feeling that if I didn't show up this time, they weren't going to invite me again…but people don't know…yes, I'm in between, going through a hard time, but that's about it. You can't admit that you're doing what you're doing. And most people don't recognize me dressed this way…I'm really careful where I go. So I basically don't go out unless I have to.

Interviewee 6: Since I can't always get what I really want, it adds to the depression…I can't go to the restaurant with my friends, "cause the guys are, like, "C'mon let me just take you for a cup of coffee or something outside"…I don't have the money to do that. I feel bad, I can't offer. A lot of times you have to go off your diet because, one, you can't afford it or, two, you have your meds. Thank God it's [diabetes medication] been covered, to compensate for what you don't have. I might go…to get a taco…or something that I haven't had in a long while, or go into the local grocery store. They have good pizza…and other things are cheap…it's also a treat…I've got a can opener that works, but how often…[do] I go…I've been eating alone in [my] apartment.

Interviewee 9: I really can't look forward to going out to eat…I'm having lunch with a friend of mine…I couldn't do that on my own…but he's treating me…I'd like to go out more, and I really can't, you know? I'm tired of saying "Well, we can't really go out…can we do this instead?" which doesn't cost any money…I feel a little bit guilty about that. …I don't know what people think of me…I can't really say "Oh I'm diabetic", and I can't do this, this, and this, without making myself look like…I'm really, really sick…and I'm not really that sick, it just means that I've got to limit myself…[if] somebody says "Let's go to a club or somewhere"…I can't drink because you know when you drink…even beer, it's got sugar in it. If I have one beer, it's going to affect me the next day…if I go to a club and order a Diet Coke, you know they're just going to say "What's wrong with him?" I'm not going to really want to get into a big explanation about how sugar affects me.

Interviewee 10: I think mostly it's the social factor of eating alone. I don't like eating alone. So…there's a community dinner that I go to, my [church group]… I've gone there for years…I go to drop-ins with my friends for two reasons: one, because I like to eat with other people, and two, because I can't afford to buy food anymore…some of the places have really good food. Some of them have food that's very high in carbohydrates, which isn't good for diabetics…plus it puts weight on you.

Interviewee 11: I go to a church and they are helpful too…if I go to church on Sunday, somebody will take me home. Most Sundays I go home with some-body…that is the next supportive group, my church…they take me home and I'm fed for the day.

Theme 3 (enhancing agency and improving resilience). Some participants considered their food insecurity to be temporary. One or more major life events, such as financial loss, job loss, marital separation, or divorce coincided with the onset of food insecurity. Many described survival strategies that buffered cur-rent adversity. For example, participants used positive self-talk, such as "Believe in your heart that it will be alright," "My life will get back on track", "You just live one day at a time", and "I will survive". Others expressed optimism through such phrases as "It's gotta be over…one way or the other".

The role of healthcare providers is pivotal in enhancing agency and resilience. Participants drew on the heartfelt help and practical diabetes management strate-gies received from healthcare providers who they found to be "genuinely caring," "gentle," and "loving" (see Interviewee 14). Feeling cared for was referred to as "the human factor" and resonated in the examples participants gave, serving as empowerment to better manage their diabetes. Other participants appreciated when their healthcare providers inquired about issues not related to diabetes (see Interviewee 15).

Participants offered suggestions to care providers on what was helpful to them: "Keep information simple for people to understand", "Listen to patients," "Be positive, and encouraging and supportive", and "Do not be judgmental". Practical advice and counselling helped them regain their agency in managing diabetes and coping with food insecurity. More specifically, participants suggested "specific inexpensive menus," "more lists, more recipes, and low-cost ideas of meals," creative ways to share food, and alternatives to food choices (see Interviewee 9). In these ways, healthcare professionals can tailor their advice specifically to the needs of food insecure clients.

Participants demonstrated resilience by implementing various strategies to better manage their diabetes, given their circumstances. For example, several participants spoke about including protein with meals, saving time and costs by batch cooking and buying starchy foods in bulk, comparison-shopping for sale items, and buying nonbrand items from discount grocery stores. Participants acknowledged that taking their medications regularly helped to compensate for their limited control over food intake and compromised food choices (see Interviewee 6 and Interviewee 2).

> Interviewee 14: I know from what I've seen with you and your staff, you people do genuinely care about us. That's a big step…it means a lot. I mean…since I was here last week till now, I've felt better about the whole situation because I know there [are] people out there that do actually care that I'm a diabetic…it really does make you feel better. It makes you wanna stay on track even more…And the way I see it, this is only human nature. Because everyone wants to feel loved somewhere, somehow, and when you know you have a group like that behind you, it, it makes you stronger.

> Interviewee 15: Every time I go see [my family doctor]…he's not just like, "Yeah, okay you're here for the visit. Thank you very much. See you next month"…He'll ask me…"Emotionally, how do you feel? Is anything bothering you? Are you getting the shakes? Are you taking too much salt in?" So…I listen to him…he's good.

> Interviewee 9: I'd say the most help I get is through this dietitian. You know, she'll tell me…"why don't you try this or you know, mix that with that?"…and, you know, I tell her my budget is limited and she'll say, "okay, this is cheaper, try that", and it'd be things I would never think of.

> Interviewee 6: Where the food is cheap…if you take your medication, it will counteract the stuff that you're not supposed to have…you eat what you can get, when you can get it. A lot of times you have to go off your diet because, one, you

can't afford it or, two, you have your meds. Thank God [they're] covered, to compensate for what you don't have.

Interviewee 2: I think just keeping a positive outlook and just…making sure I take my medications…I guess I figure that the medication will make up for some of the lapses that I've done, because I know that some people are able to perfectly manage diabetes.

9.4   DISCUSSION

Individuals living with diabetes who are food insecure face many challenges that greatly impact their ability to self-care [10–16]. Our findings demonstrate that these individuals have a limited ability to acquire, select, and prepare appropriate foods, in addition to maintaining consistent carbohydrate intake and meal spacing throughout the day. Barriers to observing an appropriate diet for diabetes management include financial constraints, a knowledge deficit for healthy meal planning on a limited budget, housing environments unconducive to food preparation and storage, inadequate community resources, and physical disabilities associated with comorbidities. The literature in Canada [10, 23] and Australia [7] confirms that many have insufficient income left after paying rent to purchase appropriate food. This was further compounded by the misperception that a "proper" diabetes diet requires "diet" foods that are low in sugar, contain artificial sweeteners, and are labelled "diabetic" [7]. Given the rising cost of healthy food [7], the circumstances of those who are food insecure can only get worse. Many study participants relied heavily on canned and convenient, high sodium, high carbohydrate foods. They stressed the inadequacy of food banks and community kitchens in meeting their special diet needs, a finding consistent with Tarasuk's research on community-based responses to food insecurity [23, 24]. López and Seligman suggest that the reliance on low-cost, energy dense foods and the inability to afford nutritious food can have a cascading effect not only on glycemic control, but also on depression, distress, and fatigue, all of which can negatively impact self-management behaviours [13]. Clinicians should therefore focus their recommendations on reducing portion size of foods that are available and accessible to them rather than focusing on food and beverage substitutions that may not be attainable [13].

Individuals with diabetes are more likely to have comorbidities than those without diabetes; it is estimated that all individuals with diabetes in Canada have some form of diabetic retinopathy, a frequent cause of legal blindness, and 40 to

50% of Canadians with type 1 or type 2 diabetes will manifest painful neuropathy within 10 years of diagnosis [4]. Several of our participants reported these complications and consequently had difficulty traveling to grocery stores, selecting fresh produce, and cooking nonprocessed foods due to sight and mobility impairments, creating a food insecure state in and of itself. Clinicians should therefore keep an inventory of meal or food delivery resources such as "Meals on Wheels," "Grocery Gateway," or "Heart to Home Meals," all of which can assist an individual with physical disabilities to access, budget for, and prepare healthy foods without leaving the home and this has been supported by other research [13].

Based on our findings, we recommend that clinicians systematically screen for food insecurity, refer to the registered dietitian as needed, tailor the care plan, and identify increased health risks, particularly for hypoglycemia that often results from missed meals or inadequate carbohydrate intake. Treatment regimens should incorporate medications that have a lower risk of hypoglycemia and that can be adapted to unpredictable or inadequate food intake as a result of food insecurity [13, 15].

Trying to manage diabetes in conjunction with a low income led to social isolation for many; most of them lived in single-person households, a common characteristic of food insecure individuals [23]. Our participants had no one to share and prepare meals with; they also limited social interactions to save costs, increasing their risk of depressive symptoms. They used resources such as community kitchens, church groups, and food coops to cope with their food insecurity and social isolation. Canadian research has shown decreased psychosocial distress and increased food security among community kitchen participants [18]. We also recommend that clinicians incorporate more group-based learning opportunities such as workshops that focus on food budgeting, low-cost meals, and, as other authors have suggested [13], strategies for more affordable healthier substitutions (e.g., buying frozen instead of fresh vegetables). Including a hands-on component such as a food skills demo is essential in engaging and motivating clients to try these strategies on their own.

Research suggests that food insecurity is cyclical, shifting between periods of food scarcity and food adequacy [14] and that food insecurity can be either chronic or temporary [25]. Similar to research among low-income Canadians, most participants in our study viewed their food insecurity as a temporary setback [10] and employed various survival strategies to endure this setback. Participants also valued the "genuine" care and "support" received from health care providers. Valued relationships with healthcare providers can be the key for patients to regain their agency to manage their diabetes during bouts of food insecurity. Norwegian research suggests that healthcare providers can encourage diabetes self-management by employing an empathetic, individualized approach [26]. It also has been suggested to move beyond the "patient-centered approach" to one that is "empowering and partnering" and that builds an emotional relationship between the healthcare professional and the client [27]. Furthermore, a recent study reported that food insecure clients respond well to diabetes self-management support programs as evidenced by a decrease in hemoglobin A1c and an increase in self-efficacy and fruit intake [28].

The study has the following limitations. While we acknowledge that the experience between those with type 1 and type 2 diabetes differs, the goal of our paper was not to compare but to obtain the general challenges that these people experience. Also, there currently is a lack of research available to distinguish the major differences in experience between the two groups. We interviewed participants in a large urban centre and, therefore, our findings may not be transferable to people in small towns and rural areas. Future research should examine the benefits of physician screening for food insecurity and diabetes self-management programs tailored to food insecure clients as this has the potential to save significant medical costs and influence the future of diabetes care.

9.5   CONCLUSION

Food insecurity presents a great challenge to diabetes self-management, an already complex chronic illness. In our study, participants faced multiple barriers to accessing appropriate foods: insufficient income, misperceptions about healthy food choices, multiple comorbidities, and inadequate cooking facilities that cumulatively impact food acquisition, selection, and preparation. Social isolation compounded these barriers, although it was somewhat buffered by the coping strategies they used and by the community food initiatives and social support networks they were able to access. Without access to healthy food, the identified barriers can potentially result in fatigue, decreased social well-being, and increased health problems, ultimately discouraging an individual to practice self-care behaviours (e.g., blood glucose monitoring, physical activity, and healthy coping). Healthcare providers should be aware of the challenges that food insecurity poses for people with diabetes, as well as the potential they have to optimize their encounters with patients. Our findings underpin the importance of understanding diabetes through the perspectives of patients' lives and tailoring diabetes management plans and community programs within the context of food insecurity.

REFERENCES

1. Canadian Diabetes Association, The Prevalence and Costs of Diabetes, Canadian Diabetes Association, Toronto, Canada, 2010.

2. P. Gæde, H. Lund-Andersen, H.-H. Parving, and O. Pedersen, "Effect of a multifactorial intervention on mortality in type 2 diabetes," New England Journal of Medicine, vol. 358, no. 6, pp. 580–591, 2008.

3. G. J. Mitchell and C. Lawson, "Living the consequence of personal choice for persons with diabetes: implications for educators and practitioners," Canadian Journal of Diabetes, vol. 24, pp. 23–30, 2000.

4. Canadian Diabetes Association, "Canadian Diabetes Association 2013 clinical practice guidelines for the prevention and management of diabetes in Canada," Canadian Journal of Diabetes, vol. 37, supplement 1, pp. S1–S212, 2013.

5. D. Lockwood, M. L. Frey, N. A. Gladish, and R. G. Hiss, "The biggest problem in diabetes," The Diabetes Educator, vol. 12, no. 1, pp. 30–33, 1986.

6. E. Gucciardi, J. A. Vogt, M. DeMelo, and D. E. Stewart, "Exploration of the relationship between household food insecurity and diabetes in Canada," Diabetes Care, vol. 32, no. 12, pp. 2218–2224, 2009.

7. B. Cuesta-Briand, S. Saggers, and A. McManus, "'You get the quickest and the cheapest stuff you can': food security issues among low-income earners living with diabetes," Australasian Medical Journal, vol. 4, no. 12, pp. 683–691, 2011.

8. H. K. Seligman, A. B. Bindman, E. Vittinghoff, A. M. Kanaya, and M. B. Kushel, "Food insecurity is associated with diabetes mellitus: results from the national health examination and nutrition examination survey (NHANES) 1999–2002," Journal of General Internal Medicine, vol. 22, no. 7, pp. 1018–1023, 2007.

9. Agriculture and Agri-Food Canada, Canada's Action Plan for Food Security: A Response to the World Food Summit, Agriculture and Agri-Food Canada, 1998.

10. F. B. Pilkington, I. Daiski, T. Bryant, M. Dinca-Panaitescu, S. Dinca-Panaitescu, and D. Raphael, "The experience of living with diabetes for low-income Canadians," Canadian Journal of Diabetes, vol. 34, no. 2, pp. 119–126, 2010.

11. N. T. Vozoris and V. S. Tarasuk, "Household food insufficiency is associated with poorer health," Journal of Nutrition, vol. 133, no. 1, pp. 120–126, 2003.

12. H. A. Bawadi, F. Ammari, D. Abu-Jamous, Y. S. Khader, S. Bataineh, and R. F. Tayyem, "Food insecurity is related to glycemic control deterioration in patients with type 2 diabetes," Clinical Nutrition, vol. 31, no. 2, pp. 250–254, 2012.

13. A. López and H. K. Seligman, "Clinical management of food-insecure individuals with diabetes," Diabetes Spectrum, vol. 25, no. 1, pp. 14–18, 2012.

14. H. K. Seligman and D. Schillinger, "Hunger and socioeconomic disparities in chronic disease," The New England Journal of Medicine, vol. 363, no. 1, pp. 6–9, 2010.

15. H. K. Seligman, T. C. Davis, D. Schillinger, and M. S. Wolf, "Food insecurity is associated with hypoglycemia and poor diabetes self-management in a low-income sample with diabetes," Journal of Health Care for the Poor and Underserved, vol. 21, no. 4, pp. 1227–1233, 2010.

16. H. K. Seligman, E. A. Jacobs, A. López, J. Tschann, and A. Fernandez, "Food insecurity and glycemic control among low-income patients with type 2 diabetes," Diabetes Care, vol. 35, no. 2, pp. 233–238, 2012.

17. D. Raphael, S. Anstice, K. Raine, K. R. McGannon, S. K. Rizvi, and V. Yu, "The social determinants of the incidence and management of type 2 diabetes mellitus: are we prepared to rethink our questions and redirect our research activities?" Leadership in Health Services, vol. 16, no. 3, pp. 10–20, 2003.

18. R. Engler-Stringer and S. Berenbaum, "Exploring food security with collective kitchens participants in three Canadian cities," Qualitative Health Research, vol. 17, no. 1, pp. 75–84, 2007.

19. United States Department of Agriculture Economic Research Service, "Household Food Security Survey Module," http://www.ers.usda.gov/topics/food-nutrition-assistance/food-security-in-the-us/survey-tools.aspx#household.

20. G. Guest, A. Bunce, and L. Johnson, "How many interviews are enough? An experiment with data saturation and variability," Field Methods, vol. 18, no. 1, pp. 59–82, 2006.

21. G. W. Ryan and H. R. Bernard, "Techniques to identify themes," Field Methods, vol. 15, no. 1, pp. 85–109, 2003.

22. J. Attride-Stirling, "Thematic networks: an analytic tool for qualitative research," Qualitative Research, vol. 1, no. 3, pp. 385–405, 2001.

23. V. Tarasuk, "A critical examination of community-based responses to household food insecurity in Canada," Health Education and Behavior, vol. 28, no. 4, pp. 487–499, 2001.

24. S. I. Kirkpatrick and V. Tarasuk, "Food insecurity and participation in community food programs among low-income Toronto families," Canadian Journal of Public Health, vol. 100, no. 2, pp. 135–139, 2009.

25. V. Tarasuk, Discussion Paper on Household Food Insecurity, Health Canada, Ottawa, Canada, 2010.

26. B. Oftedal, B. Karlsen, and E. Bru, "Perceived support from healthcare practitioners among adults with type 2 diabetes," Journal of Advanced Nursing, vol. 66, no. 7, pp. 1500–1509, 2010.

27. C. L. McWilliam, "Patients, persons or partners? Involving those with chronic disease in their care," Chronic Illness, vol. 5, no. 4, pp. 277–292, 2009.

28. C. R. Lyles, M. S. Wolf, D. Schillinger et al., "Food insecurity in relation to changes in hemoglobin A1c, self-efficacy, and fruit/vegetable intake during a diabetes educational intervention," Diabetes Care, vol. 36, no. 6, pp. 1448–1453, 2013.

Children's Very Low Food Security is Associated with Increased Dietary Intakes in Energy, Fat, and Added Sugar among Mexican-Origin Children (6-11 Y) in Texas Border *Colonias*

Joseph R. Sharkey, Courtney Nalty,
Cassandra M. Johnson, and Wesley R. Dean

10.1 INTRODUCTION

The Southwestern United States border region is home to many *colonias*. These settlements are occupied by a growing population of people who share a similar Mexican heritage, language, and socioeconomic standing and who have unacceptably high rates of poverty, adult and childhood obesity, and food insecurity [1–3]. Border-region *colonias* can be considered an archetype for the increasing number of new destination immigrant communities [1]. Many of these communities of Mexican immigrants are located throughout the United States, including many non-traditional interior locales [4–6].

Food insecurity underpins an emerging national issue of nutritional health inequity among Mexican-origin and Hispanic households. The 2009 Current

© *Sharkey et al; licensee BioMed Central Ltd. 2012; BMC Pediatrics 2012, 12:16; DOI: 10.1186/1471-2431-12-16. Distributed under the terms of the Creative Commons Attribution License (http://creativecommons.org/licenses/by/2.0/).*

Population Survey Food Security Supplement identified low or very low food security in 26.9% of Hispanic households, compared with 14.7% in all U.S. households, and in 18.7% of Hispanic households with food-insecure children compared with 10.6% in all households [7]. In a study of Mexican immigrant families in Minnesota, Kersey and colleagues observed much higher rates of child hunger among 1,310 Mexican immigrant families than among 1,805 non-immigrant families (6.8% versus 0.5%) [8]. Food insecurity is much more prevalent among Mexican-origin households in the Texas border region compared with other regions of the U.S [3, 7]. In a study of food access among 610 adult women in Texas border *colonias*, researchers found 49% of all households and 61.8% of households with children could be classified at the most severe level of food insecurity—child insecure [3].

At the same time, nutrition-related health conditions, such as obesity and type 2 diabetes, are increasing in Mexican-origin youth. Risk factors for obesity and type 2 diabetes are more common in Mexican-origin children than other racial/ethnic groups [9–11] and include increased intakes of energy-dense and nutrient-poor foods, such as fats, sweeteners, desserts, and salty snacks. Energy-dense foods are highly palatable and promote higher calorie intakes [12, 13]. Diets with proportionally more contribution from energy-dense foods increase the risk for inadequate intakes of vitamin D, calcium, potassium, and dietary fiber and the likelihood of consuming excessive amounts of added sugar, fats, and sodium [14]. For limited-resource populations (households with limited economic and physical resources and limited access to healthy foods), including children of Mexican-origin, energy-dense foods may also be more accessible, available, and affordable [2, 15–17]. This may be especially true for Mexican-origin children in Texas border colonias who reside in food-insecure households in communities lacking access to nutritious food [2, 3, 18]. For children, residing in a food-insecure household can prevent them from achieving the nutrient intake needed for optimal development and health, as well as impede their academic performance [19–26].

Thus, it is critically important to understand the relationship between food insecurity and children's dietary intake among limited-resource Mexican-origin children. However, few studies have examined this association. Prior studies among Hispanics relied exclusively on parental reports of household food-supply adequacy and their child's diet and experiences. These studies revealed multiple associations between food insecurity and diet. Food-insecure children were less likely to meet recommended food-group guidelines [27], have greater intakes of fats, saturated fats, sweets, and fried foods [28], and lower fruit and vegetable intake at home [29]. Among 5th grade students who reported dietary intake using three 24-hour dietary recalls and whose mothers reported food security, food-security status was not associated with dietary intake [30]. Only one study assessed dietary intake through child-reported dietary recalls, and none measured food security from the child's perspective.

Although mothers often spare their children from nutritional deprivation and report that children are more protected from household food insecurity [31, 32], this experience is from the parent's perspective [33]. There has been a call for research to assess the relationship between food security and children's diet [34], yet little research has focused on child's perceptions or experiences of food insecurity and their association with dietary intake [35]. Current measures represent household food security status of the household or children within the household as a group, rather than the experiences of a particular individual within the household [36]. Children best report their own experiences [37]. Measurements of food security as reported by the child, which may be more sensitive of the food issues experienced by children, is essential for understanding the influence of food insecurity on the nutritional health of children [35]. Understanding the relationship between children's experience of food security and their dietary intake [30] is needed to comprehend the effect of food insecurity on children's nutrient intake [38]. The current study seeks to assess the relationship between children's experience of food insecurity and nutrient intake from food and beverages by (1) assessing food security status as reported by 50 Mexican-origin children (ages 6-11 years), (2) examining nutrient intakes from three 24 hour dietary recalls from each child, and (3) determining the relationship between food security status and nutrient intake.

10.2 METHODS

10.2.1 Setting

The study was conducted in two large areas of *colonias* in Hidalgo County, located in the Lower Rio Grande Valley of Texas along the Mexico border. From prior work and in consultation with community partners [2], 10 census block groups (CBG) were identified in the western part of the county and 10 in the eastern portion of the county. In both areas, a majority of CBGs are considered to be highly deprived neighborhoods [2]. Highly deprived neighborhoods are those with overall high proportions of unemployed adults, households without telephone service, families receiving public assistance, households lacking complete kitchen facilities, households lacking complete plumbing facilities, adults with less than 10 years of education, or those living below the poverty threshold [2]. Forty *colonias* were spatially selected, with at least one colonia in each of the 20 CBGs.

10.2.2   Study Sample

The study sample consisted of 50 family dyads (mother and child 6-11 years), who were recruited for a cohort study by team *promotora*-researchers; 25 dyads were recruited from western area *colonias* and 25 from eastern area *colonias*. Letters of invitation were personally delivered by *promotora*-researchers, and eligibility was determined by the presence of one child (age 6-11 years) residing in the household. The study was explained to each prospective adult participant (e.g., inform about assessments, confidentiality, financial incentive, etc.), and the first of three in-home assessments was scheduled within two days. The mother provided consent for both members of the dyad to participate in the study, and the child provided assent for participation. All materials and protocols were approved by the Texas A&M University Institutional Review Board.

10.2.3   Data Collection

This analysis focuses on data collected March to June 2010 from all 50 children during three in-home visits: survey data and anthropometrics from the first visit and dietary recalls from all three visits. The survey included sections on demographics and food security and was interviewer-administered by promotora-researchers, who received training in collection of survey, anthropometric measures, and 24-hour dietary recalls. All materials were reviewed by community partners and were validated by local/area experts. A pilot test was conducted in colonias not selected from the study area and necessary modifications were made. Promotora-researchers received the equivalent of four full days of training on data collection and protection of participant confidentiality. All measures were translated into Spanish using translation-back translation method with the following steps: 1) translation of the original English into Spanish, ensuring that the English meaning is maintained; 2) back-translation into English by an independent translator who is blinded and is not familiar with either the Spanish or English version; 3) comparison of the two English versions; and 4) resolution of any discrepancies. Community partners and pro-motora-researchers verified translation accuracy and appropriateness to ensure semantic, conceptual, and normative equivalence. All survey and 24-hour dietary recall data were collected in Spanish, which was the language spoken in the homes of all participants.

10.2.4   Measures

Demographics included child's sex, age, school grade, and country of birth.

Anthropometric measure of body mass index (BMI) was used to gain a general sense of body fatness. Weight was measured (to the nearest 0.1 kg) in the home with a portable, self-zeroing scale. Weight was measured twice, with the children wearing light clothing and no shoes. Standing height (to the nearest mm) was measured twice with a portable stadiometer. Using the mean of the two measures of weight and height, BMI was calculated as weight (kg)/[height (m)]^2. Appropriate Centers for Disease Control and Prevention (CDC) BMI-for-age-and-sex growth charts were used to classify each child's BMI status as underweight (< 5th percentile), healthy weight (5th percentile to < 85th percentile), overweight (85th percentile to < 95th percentile), or obese ($\geq$95th percentile) [39, 40].

Children's food security was assessed using the nine-item child food security measure developed by Connell and colleagues [33]. Pilot testing of the food security measure with a sample of children similar to participant children was performed to determine understandability and face validity. Participant children were asked by the *promotora*-researcher whether they experienced each of the nine items during the last three months (see Table 1). Response options included "a lot", "sometimes," or "never." Each item was constructed as a binary variable; yes (a lot or sometimes) vs. no (never). Iterative common factors analysis on the nine items identified one factor (eigenvalue = 3.2) that explained 79% of the shared variance; and internal reliability was good (Cronbach's α = 0.81). Affirmative responses to the nine items were summed into an ordinal children's food security score [34], which was used to categorize each child as having high food security (score = 0), marginal food security (score = 1), low food security (score = 2-4), or very low food security (score = 5-9).

TABLE 10.1 Children self-reported food security (n = 50)

In the last 3 months,	N (%)[1]
1. Did you worry that food at home would run out before your family got money to buy more?	25 (50)
2. Did the food that your family bought run out and your family did not have money to get more?	23 (46)
3. Were you not able to eat a variety of healthy foods at a meal because your family didn't have enough money?	20 (40)
4. Did your meals only include a few kinds of cheap foods because your family was running out of money to buy food?	27 (54)
5. Was the size of your meals cut because your family didn't have enough money for food?	17 (34)
6. Did you have to eat less because your family didn't have enough money to buy food?	16 (32)
7. Did you have to skip a meal because your family didn't have enough money for food?	15 (30)
8. Were you hungry but didn't eat because your family didn't have enough food?	4 (8)
9. Did you not eat for a whole day because your family didn't have enough money for food?	6 (12)
Food Security Categories	
High food security	9 (18)
Marginal food security	9 (18)
Low food security	18 (36)
Very low food security	14 (28)

[1]Affirmative response = combination of "a lot" and "sometimes"

10.2.5 Dietary Intake

Three 24-hour (previous day) dietary recalls occurring on randomly selected, nonconsecutive days (one recall measured weekend intake and two measured weekday intake) were collected in the home from each participant child by the same interviewer (*promotora*-researcher). In most cases, the mother observed and assisted the children if necessary. Dietary intake training for the interviewers included review of all protocols and scripts, modeling of interviewing, practice interviews with children similar in age to the study participants, use of tools for portion-size estimation, quality control, and focus on children's reporting of food items. The first recall occurred during the first in-home visit, and the second and third recalls were collected in the home during the second and third visits (within two weeks of the first visit). Detailed information on food and beverage consumption, including description, brand name, location of preparation and consumption, and preparation method during the previous day was collected using standardized protocols following a modification of the multiple-pass interview technique of the Nutrition Data System for Research (NDS-R) [41]. Data were collected on hard copy in Spanish, modified from an approach previously used [42], and then entered into NDS-R 2009 in English [41]. Children were first asked to provide a quick list of generic food and beverage items consumed during the previous day based on short time intervals (e.g., before breakfast, at breakfast, between breakfast and lunch, and at dinner); prompts included food consumption occasions and locations. This was followed by a review of the quick list. During this pass, the interviewer probed for forgotten foods by asking about snacks and beverages (including water) and about the source of the food or beverage. The third pass provided food details such as the time and place of the eating occasion, food descriptions, brand name, ingredients and preparation, and portion size and quantity consumed. As a result of pilot testing, multiple approaches were used for estimation of portion size and included measurement of typically-used cups, glasses, bowls, and containers in the home, food and beverage models, geometric shapes (circles, rectangles, and wedges), and three-dimensional thickness aides. The fourth pass was a final and comprehensive review of the previous-day's intake. Nutrient calculations were performed using NDS-R 2009 software. Three-day mean nutrient intakes, with equal weighting for each of the three days (2 weekdays and 1 weekend) of dietary recall were calculated for each child for total energy (kcal), protein (g), dietary fiber (g), calcium (mg), vitamin D (mcg), potassium (mg), sodium (mg), Vitamin C (mg), percentage of calories from fat, percentage of calories from added sugars, and percentage of calories from saturated fat.

10.2.6   Analysis

All analyses were performed using Stata Statistical Software: Release 11 (College Station, TX: StataCorp, 2009). Descriptive statistics were calculated for each child's baseline characteristics, BMI status, food security status, and nutrient intake. Wilcoxon Signed-Rank Test was used to compare weekend and weekday nutrient intake by level of food security. Non-parametric test for trend across ordered groups of food security was used to examine each nutrient. Separate multiple regression models with robust (White-corrected) Standard errors (SEs), were individually fitted for total energy (kcal), protein (g), dietary fiber (g), calcium (mg), vitamin D (mcg), potassium (mg), sodium (mg), Vitamin C (mg), percentage of calories from fat, percentage of calories from added sugars, and percentage of calories from saturated fat. All models included sex, age, country of birth, BMI status, and food security status as independent variables. These variables were selected based on their documented association with dietary intake.

10.3 RESULTS

Table 2 shows baseline characteristics for the 50 participant children. Sixteen children (32%) were born in Mexico and twenty-one (42%) were overweight or obese based on the Centers for Disease Control and Prevention BMI-for-age-and-sex growth charts [40]. Results from the nine-item children's food security measure are shown in Table 1. Thirty-two children (64%) reported low (n = 18) or very low food security (n = 14). Although BMI status was not significantly associated with food-security status, 48 percent of the sample was measured as being of healthy weight and reporting low or very low food security. Nutrient intake and dietary recommendations for the entire sample and nutrient intake by food-security status are shown in Table 3. Using the 2010 Institute of Medicine age- and sex-specific recommendations [43], 28% (n = 14) met the recommendations for calcium, none for potassium or vitamin D, 10% (n = 5) for dietary fiber, and 6% (n = 3) for sodium (data not shown). Although all children exceeded the recommendation for protein, as a percent of total calories, protein intake ranged from 11.8% to 22% (data not shown). Weekend intakes for calcium, vitamin D, potassium, and vitamin C were significantly lower than weekday consumption, and percentage of calories from fat, and combined percentage from fat and added sugar (data not shown) were significantly higher on weekends than weekdays. Children who were identified with low food security consumed significantly less calcium and vitamin D on weekends, compared with weekdays, and very low food-security children consumed a greater percentage of calories from fat on weekends than weekdays. Three-day average intake for total energy, protein, percentage of calories from added sugar, and percentage of calories from saturated fat demonstrated a significant and positive trend (indicating greater intake) with reduced food-security status. The same positive trend was observed for weekend intake of total energy, and for weekday intake of percentage of calories from added sugars and saturated fat.

TABLE 10.2   Baseline characteristics of Mexican-origin children (n = 50)

	Mean ± SD (Median)	N (%)
Sex		
Female		31 (62)
Age, y	9.1 ± 1.3 (9.2)	
Country of birth		
Mexico		16 (32)
United States		34 (68)
Weight status[a]		
Underweight		5 (10)
Healthy weight		24 (48)
Overweight		9 (18)
Obese		12 (24)

[a] Based on BMI-for-age-and-sex growth charts. Underweight = < 5th percentile; healthy weight = 5th percentile to < 85th percentile; overweight = 85th percentile to < 95th percentile; and obese = ≥ 95th percentile

The multiple regression results, presented in Table 4, show the association of children's food-security status, sex, age, BMI status, and country of birth to nutrient-specific intakes. Very low food security was associated with greater intakes of total energy, calcium, and percentage of calories from added sugar. In data not shown in Table 4, marginal ($\beta = 4.8$, SE = 2.2, p = 0.032), low ($\beta = 4.4$, SE = 1.9, p = 0.028), and very low ($\beta = 8.4$, SE = 2.0, p < 0.001) food security were associated with increased intake as a percentage of calories from combined fat and added sugar. In addition, increased age was associated with lower intakes of calcium and vitamin D; being born in Mexico with greater intake of sodium; and being female with lower calcium intake. BMI status was not associated with any of the nutrients.

10.4   DISCUSSION

Previous work has recognized the relationship between food security and children's diets [34, 35, 37], but this is apparently the first study to assess the relationship between food security and children's nutrient intakes in Mexican-origin children, based on children's reports of both their experiences of food

TABLE 10.3   Nutrient intake (3-day average, weekday average, and weekend) overall and by food security status[a]

	Dietary Recommendations		Total Sample (N = 50)	Food secure (N = 9)	Marginal food secure (N = 9)	Low food security (N = 18)	Very low food security (N = 14)
	6-8 y	9-11 y					
Total energy (kcal)	1400-1600	1600-2200					
3-day average[c2]			2034.9	1766.9	2049.2	1994.2	2250.2
			(372.8)	(354.5)	(287.0)	(340.7)	(376.6)
Weekday			2008.1	1757.4	2081.7	2020.7	2105.8
			(375.8)	(313.9)	(318.0)	(363.4)	(423.0)
Weekend day[c2]			2097.5	1785.9	2023.0	1947.0	2539.1
			(717.6)	(761.6)	(687.5)	(611.9)	(700.9)
Protein (g)	19	34					
3-day average[c1]			82.9	73.3	78.4	84.8	89.5
			(19.7)	(17.9)	(14.1)	(20.2)	(22.0)
Weekday			81.1	70.7	78.8	86.7	82.1
			(19.9)	(18.0)	(11.8)	(21.7)	(21.9)
Weekend day			86.4	78.6	76.5	81.6	104.1
			(36.7)	(36.3)	(30.2)	(35.8)	(39.3)
Dietary fiber (g)	17-20	22-25					
3-day average			15.2	15.7	14.3	16.2	14.1
			(5.0)	(3.3)	(4.2)	(5.3)	(6.2)
Weekday			15.5	16.6	16.2	16.5	13.1
			(5.3)	(5.1)	(5.6)	(5.8)	(4.3)
Weekend day			14.6	13.8	10.8	15.6	16.1
			(9.8)	(6.9)	(4.6)[b1]	(7.1)	(15.4)

TABLE 10.3 (*Continued*)

	Dietary Recommendations		Total Sample	Food secure	Marginal food secure	Low food security	Very low food security
	6-8 y	**9-11 y**	**(N = 50)**	**(N = 9)**	**(N = 9)**	**(N = 18)**	**(N = 14)**
Calcium (mg)	1000	1300					
3-day average			993.4	820.4	1077.5	951.6	1104.3 (387.6)
			(300.5)	(266.4)	(245.4)	(228.1)	
Weekday			1056.3	918.2	1131.5	1095.7	1046.2
			(299.6)	(360.2)	(149.1)	(321.3)	(300.8)
Weekend day			898.2	624.7	999.7	733.6	1220.6
			(566.2)[b2]	(355.8)	(609.7)	(381.5)[b2]	(709.3)
Vitamin D (mcg)	15	15					
3-day average			6.9	5.4	7.6	6.2	8.4
			(2.9)	(2.9)	(1.9)	(2.5)	(3.4)
Weekday			7.5	6.3	7.9	7.2	8.6
			(2.7)	(3.1)	(2.6)	(2.2)	(3.0)
Weekend day			5.8	3.6	6.8	4.6	8.0
			(4.4)[b3]	(3.5)[b1]	(2.1)	(4.3)[b2]	(5.1)
Potassium (mg)	3800	4500					
3-day average			2530.1	2499.1	2489.6	2457.6	2669.4
			(554.4)	(482.4)	(451.5)	(530.4)	(701.4)
Weekday			2640.3	2649.6	2692.4	2650.3	2588.0
			(602.8)	(573.2)	(466.5)	(695.9)	(628.9)
Weekend day			2332.9	2197.9	2052.4	2152.2	2832.3
			(955.2)[b1]	(976.7)	(716.3)	(782.3)	(1164.0)

TABLE 10.3 (Continued)

	Dietary Recommendations		Total Sample	Food secure	Marginal food secure	Low food security	Very low food security
	6-8 y	9-11 y	(N = 50)	(N = 9)	(N = 9)	(N = 18)	(N = 14)
Sodium (mg)	< 1900	< 2200					
3-day average			3450.7	3083.0	3165.0	3687.8	3565.9
			(913.2)	(750.9)	(330.6)	(1050.7)	(1021.4)
Weekday			3372.1	3100.0	3180.7	3698.0	3251.0
			(994.8)	(881.8)	(357.7)	(1227.5)	(981.2)
Weekend day			3553.5	3049.0	3042.9	3561.5	4195.7
			(1695.0)	(1074)	(1198.7)	(1934.2)	(1881.8)
Vitamin C (mg)	25	45					
3-day average			101.1	114.2	92.7	87.8	113.7
			(52.7)	(64.6)	(55.9)	(36.8)	(59.4)
Weekday			110.5	124.7	105.1	93.8	122.6
			(63.8)	(72.2)	(69.2)	(57.4)	(64.2)
Weekend day			81.4	93.3	62.2	78.9	95.8
			(70.4)[b2]	(98.6)	(56.8)	(52.8)	(82.0)
Fat (percent of calories)	25-35	25-35					
3-day average			34.0	32.2	33.0	35.3	34.2
			(4.1)	(4.4)	(3.8)	(4.5)	(3.2)
Weekday			32.5	30.3	32.9	33.4	32.7
			(4.2)	(3.2)	(5.3)	(4.4)	(3.7)
Weekend day			36.8	35.9	33.6	38.7	37.1
			(7.8)[b3]	(10.1)	(4.0)	(9.7)	(4.9)[b1]

TABLE 10.3 (Continued)

	Dietary Recommendations		Total Sample	Food secure	Marginal food secure	Low food security	Very low food security
	6-8 y	9-11 y	(N = 50)	(N = 9)	(N = 9)	(N = 18)	(N = 14)
Added sugars (percent of calories)							
3-day average[c1]			15.0	11.8	16.5	13.0	18.7
			(5.5)	(2.8)	(5.3)	(4.5)	(6.0)
Weekday[c1]			14.8	11.1	15.3	13.5	18.7
			(6.3)	(4.1)	(5.2)	(6.0)	(7.0)
Weekend day			15.5	13.5	20.0	12.1	18.3
			(7.7)	(4.7)	(9.0)	(6.7)	(7.9)
Saturated fat (percent of calories)							
3-day average	< 10	< 10	12.6	11.4	12.8	12.8	12.8
			(2.0)	(2.0)	(1.4)	(2.2)	(1.9)
Weekday			12.3	11.4	13.1	12.5	12.0
			(2.2)	(1.7)	(2.3)	(2.4)	(2.2)
Weekend day			13.2	11.4	13.1	13.4	14.2
			(4.2)	(3.8)	(3.8)	(4.7)	(3.8)

Dietary Recommendations for children from USDA/ARS Children's Nutrition Research Center at Baylor College of Medicine, available at http://www.bcm.edu/cnrc/consumer/archives/percentDV.htm

[a] Nutrient intake reported as mean (SD)

[b] Comparison of weekday and weekend day nutrient intake by level of food security, using the Wilcoxon Signed-Rank Test

[c] Test for trend across ordered groups of food security

Level of statistical significance: [1] p < 0.05 [2] p < 0.01 [3] p < 0.001

TABLE 10.4 Children's food security and demographic correlates of children's nutrient intakes from multiple regression models[1]

	Energy (kcal)	Protein (g)	Fiber (g)	Calcium (mg)	Vitamin D (mcg)	Potassium (mg)	Sodium (mg)	Vitamin C (mg)	Fat (% kcal)	Added Sugar (% kcal)
Food security										
Marginal	238.06	2.34	-0.65	270.57	1.93	-96.78	-62.01	-16.45	0.16	4.15
	(142.06)	(8.16)	(2.21)	(109.28)**	(1.24)	(235.54)	(288.16)	(25.06)	(2.22)	(2.14)
Low	166.71	8.54	0.16	91.00	0.60	-109.31	377.84	-30.17	2.25	1.65
	(140.92)	(7.39)	(2.02)	(102.56)	(1.09)	(190.62)	(371.94)	(21.19)	(1.77)	(1.48)
Very low	377.16	10.54	-0.94	187.29	1.71	22.11	278.73	-3.68	0.97	6.82
	(169.91)*	(9.32)	(2.30)	(139.25)*	(1.40)	(275.27)	(365.86)	(25.64)	(1.80)	(2.05)***
Female	-118.83	-5.51	-1.20	-189.29	-0.86	-130.99	-124.02	0.66	-1.44	2.31
	(124.52)	(6.87)	(1.89)	(84.62)*	(0.82)	(194.59)	(362.36)	(19.27)	(1.40)	(1.56)
Age	-48.27	-2.24	0.89	-83.56	-1.18	-67.15	-30.30	-6.74	-0.35	0.04
	(54.08)	(2.58)	(0.69)	(29.02)**	(0.28)***	(76.77)	(115.29)	(6.44)	(0.45)	(0.54)
BMI status										
Overweight	-60.29	-4.65	0.65	50.17	0.42	-183.03	7.43	-3.19	-0.85	0.12
	(150.37)	(8.32)	(2.16)	(79.68)	(0.77)	(210.69)	(440.01)	(21.13)	(1.77)	(1.62)
Obese	22.25	2.09	0.35	50.11	1.05	-11.24	-81.29	-21.66	-1.87	2.02
	(121.58)	(7.37)	(1.44)	(109.81)	(1.04)	(202.18)	(300.96)	(17.85)	(1.64)	(1.68)
Country of birth										
Mexico	78.80	3.73	1.66	-53.06	-0.85	43.00	689.84	-10.73	0.76	0.23
	(118.59)	(6.49)	(2.13)	(80.62)	(0.59)	(211.25)	(337.07)*	(20.01)	(1.44)	(1.73)
R^2	0.271	0.156	0.120	0.353	0.478	0.078	0.220	0.118	0.145	0.290

[1] Coefficients are reported; robust SEs are in parenthesis and are corrected with the White-Huber correction. There were 50 observations. Reference categories (variables) omitted to prevent perfect collinearity: food security (food secure), female (male), BMI status (Normal/Underweight), and country of birth (U.S.). Age was entered as a continuous variable.

*$p < 0.05$ **$p < 0.01$ ***$p < 0.001$

insecurity and dietary intake. The national prevalence of household food insecurity is greater among Hispanic households in the U.S [7] and substantially greater among Mexican-origin households in *colonias* [3]. Findings from this study expand our understanding of the experience of food insecurity by school-age, Mexican-origin children and the association of food-security status with nutrient intakes.

Results present additional evidence that food insecurity is more prevalent among Mexican-origin children in Texas border *colonias* than previous estimates suggested. For instance, national data from 2009 indicated that 18.7% of Hispanic households, regardless of race or country of origin, had food-insecure children [7]. A community-based nutrition assessment of 610 Mexican-origin adults in Texas border *colonias* reported that 49% of all households and 61.8% of households with children were classified as child food insecure [3]. In the current sample, 28% of children reported very low food security and 64% reported low or very low food security. At least one-third of children reported having to skip a meal, go hungry, or not eat for a whole day because of limited or no food resources in the home, which supports and is supported by the community assessment. Children's total energy and nutrient intakes in this study were similar to or greater than previously reported among Mexican-American children (6-11 years) [44]. In addition, the present study showed that decreasing food-security status was associated with increased intake of total calories and percentage of calories from fat and added sugars, which confirms the work of Rosas and colleagues [28], and is in contrast to the work of Matheson and colleagues, which reported a not-significant relationship between household food security and children's dietary intake [30].

This paper not only emphasizes the alarming rates of food insecurity for this Hispanic subgroup, but describes the associations for food insecurity and diet among this sample of Mexican-origin children. Such findings have implications at a regional and national level, as the Mexican-origin population continues to grow along the border and in new destination communities [8, 45]. Immigrants from Latin America have provided the largest percent of foreign-born population since 1990: 44.3% in 1990, 51.7% in 2000, and 53.6% in 2007 [46]. People of Mexican-origin represent approximately 64% of both the native and foreign-born Hispanic population [46]. Of the 29.2 million Mexican Hispanics, 40 percent were foreign born. The percentage of all children living in the United States with at least one foreign-born parent increased from 15 percent in 1994 to 23 percent in 2010 [47]. In 2010, 33 percent of foreign-born children with foreign-born parents and 26 percent of native children with foreign-born parents lived below the poverty line [47]. Data presented here may foreshadow higher rates of nutrition-related health conditions, such as obesity, type 2 diabetes, and cardiovascular disease among Mexican-origin children. Child-reported food insecurity situations could serve as a screen for nutrition problems in children. Further, the National School Lunch and School Breakfast Programs, which play a major beneficial role in children's weekday intakes, may not be enough to keep pace with the nutritional needs of low and very low food secure Mexican-origin children.

The present study has several particular strengths. First, this is a study of hard-to-reach Mexican-origin children in border *colonias*. This population is of increasing national importance because such *colonias* can be considered an archetype for the new-destination Mexican immigrant communities that are now found in great numbers throughout the U.S. Second, to our knowledge, this is the first study that uses children's report of their food-insecurity experiences in the past three months to describe food-security status, which is preferable and reduces the cognitive burden placed on respondents by the conventional twelve-month time period [7]. As such, this study builds on the work of Connell and colleagues and Fram and colleagues [33, 35, 37], that identified the importance of the child's perspective in understanding food insecurity and its consequences [35, 37]. Third, usual dietary intake was determined by three 24-hour dietary recalls that included weekday and weekend intakes. Each recall was conducted face-to-face in the home, multiple strategies were used to estimate portion size, and a modified multiple-pass method was used to capture home and away-from-home (e.g., school) food intake. Young children can provide information on their diet as accurately, or more accurately, than their parents, especially for food eaten outside the home [48–51].

There are several limitations to this study that warrant mention. Data were collected during one season of the year, which limits our ability to describe seasonal variation in dietary intake or food-security status or to make causal inferences. This could have important implications for times of year when children are unable to participate in school nutrition programs, such as during the summer or holidays. Although the three-month time frame was much better than asking about food security experiences in the last 12 months, this study did not collect data on frequency or duration of food insecurity situations. This limits our ability to differentiate between acute and chronic food insecurity. An additional limitation is an absence of data on food coping strategies employed by children to help manage food resources [37]. Finally, the study sample is small and is limited to two areas of *colonias* in the Texas border region, which limits our ability to generalize these results. Future work should focus on expanding our understanding of seasonal variation in the frequency and duration of children's experiences of food insecurity.

10.5   CONCLUSIONS

Despite these limitations, the results of this study further the knowledge of children's experiences of low and very low food security and the association of food security status with children's dietary intake. The Mexican-origin population is rapidly expanding throughout the United States; record numbers of individuals and families are experiencing food insecurity, and for children living in rural or underserved areas such as the colonias, food insecurity may be an ongoing reality. The prevalence of low and very low food security in this border area is alarming, despite the participation of all study participants in the School Breakfast and National School Lunch Programs. The high prevalence of low and very low food security among these children is especially troubling given the importance of good nutrition on optimal growth, function, and health [19, 20]. Young children of Mexican immigrant families have a greater risk for hunger and household food insecurity [8], and are less likely to meet dietary recommendations than other children [27]. In this sample of Mexican-origin children, not only did most of the children not meet dietary recommendations in key nutrients, but children with very low food security consumed higher levels of energy, fat, and added sugars. The results also indicate the importance of further examining the frequency and duration of low and very low food security in children. Enhanced research efforts are needed that will lead to better understanding of coping strategies and the use of federal and community food and nutrition assistance programs for reducing food insecurity. Clearly, systematic and sustained action on multiple levels that integrates multi-sector partnerships and networks is needed for culturally-tailored health promotion and policy efforts to reduce child food insecurity.

REFERENCES

1. Esparza AX, Donelson AJ: The colonias reader: economy, housing and public health in U.S.-Mexico. 2010, Tucson: The University of Arizona Press
2. Sharkey JR, Horel S, Han D, Huber JC: Association between Neighborhood Need and Spatial Access to Food Stores and Fast Food Restaurants in Neighborhoods of Colonias. Int J Health Geogr. 2009, 8: 9-10.1186/1476-072X-8-9.

3. Sharkey JR, Dean WR, Johnson CM: Association of household and community characteristics with adult and child food insecurity among mexican-origin households in Colonia along the Texas-Mexico border. Int J Equity Health. 2011, 10: 19-10.1186/1475-9276-10-19.

4. Zúñiga V, Hernández-León R: New destinations: mexican immigration in the United States. 2006, New York: Russell Sage Foundation

5. Bernosky de Flores CH: Human capital, resources, and healthy childbearing for Mexican women in a new destination immigrant community. J Transcult Nurs. 2010, 21 (4): 332-341. 10.1177/1043659609360714.

6. Cornfield DB: Immigrant labor organizing in a "new destination cirty": approaches to the unionization of African, Asian, Latino, and Middle Eastern workers in Nashville. Global Connection & Local Receptions: New Latino Immigration to the Southeastern United States. Edited by: Ansley F, Shefner J. 2009, Knoxville, TN: University of Tennessee Press

7. Nord M, Coleman-Jensen A, Andrews M, Carlson S: Household Food Security in the United States, 2009. vol ERR-108. 2010, Washington: U.S. Department of Agriculture, Economic Research Service

8. Kersey M, Geppert J, Cutts DB: Hunger in young children of Mexican immigrant families. Public Health Nutr. 2007, 10 (4): 390-395.

9. Treviño RP, Marshall RM, Hale DE, Rodriguez R, Baker G, Gomez J: Diabetes risk factors in low-income Mexican-American children. Diabetes Care. 1999, 22: 202-207. 10.2337/diacare.22.2.202.

10. Ogden C, Carroll M: Prevalence of obesity among children and adolescents: United States, trends 1963-1965 through 2007-2008. NCHS Health E-Stat Hyattsville. 2010, MD: US Department of Health and Human Services, CDC, National Center for Health Statistics, (Last Accessed 18 October 2010), [http://www.cdc.gov/nchs/data/hestat/obesity_child_07_08/obesity_child_07_08.htm]

11. Wang Y, Beydoun MA: The obesity epidemic in the United States - gender, age, socioeconomic, racial/ethnic, and geographic characteristics: a systematic review and meta-regression analysis. Epidemiol Rev. 2007, 29: 6-28. 10.1093/epirev/mxm007.

12. Kant AK: Consumption of energy-dense, nutrient-poor foods by adult Americans: nutritional and health implications. The third national health and nutrition examination survey, 1988-1994. Am J Clin Nutr. 2000, 72: 929-936.

13. Agency for Healthcare Research and Quality: Diabetes Disparities Among Racial and Ethnic Minorities. [http://www.ahrq.gov/research/diabdisp.htm]

14. Dietary Guidelines Advisory Committee: The Report of the Dietary Guidelines Advisory Committee on the Dietary Guidelines for Americans, 2010. 2010, Washington, DC: Department of Health and Human Services

15. Drewnowski A, Specter S: Poverty and obesity: the role of energy density and energy costs. Am J Clin Nutr. 2004, 79: 6-16.

16. Drewnowski A: Fat and sugar: an economic analysis. J Nutr. 2003, 133: 838S-840S.

17. Monsivais P, Aggarwal A, Drewnowski A: Following federal guidelines to increase nutrient consumption May lead to higher food costs for consumers. Health Aff. 2011, 30 (8): 1471-1477. 10.1377/hlthaff.2010.1273.

18. Nord M: Food Insecurity in Households with Children: Prevalence, Severity, and Household Characteristics. 2009, United States Department of Agriculture, Economic Research Service

19. Alaimo K, Olson CM, Frongillo EA: Food Insufficiency and American school-aged children's cognitive, academic, and psychosocial development. Pediatrics. 2001, 108 (1): 44-53.
20. Jyoti DF, Frongillo EA, Jones SJ: Food security affects school Children's academic performance, weight gain, and social skills. J Nutr. 2005, 135: 2831-2839.
21. Weinreb L, Wehler C, Perloff J, Scott R, Hosmer D, Sagor L, Gundersen C: Hunger: its impact on Children's health and mental health. Pediatrics. 2002, 110 (4): e41-10.1542/peds.110.4.e41.
22. Marjerrison S, Cummings EA, Glanville NT, Kirk SFL, Ledwell M: Prevalence and associations of food insecurity in children with diabetes mellitus. J Pediatr. 2011, 158: 607-611. 10.1016/j.jpeds.2010.10.003.
23. Lohman BJ, Stewart S, Gundersen C, Garasky S, Eisenmann JC: Adolescent overweight and obesity: links to food insecurity and individual, maternal, and family stressors. J Adolesc Heal. 2009, 45: 230-237. 10.1016/j.jadohealth.2009.01.003.
24. Gundersen C, Kreider B: Bounding the effects of food insecurity on children's health outcomes. J Heal Econ. 2009, 28: 971-983. 10.1016/j.jhealeco.2009.06.012.
25. Yoo JP, Slack KS, Holl JL: Material hardship and the physical health of school-aged children in Low-income households. Am J Public Health. 2009, 99: 829-836. 10.2105/AJPH.2007.119776.
26. Hamelin A-M, Habicht J-P, Beaudry M: Food insecurity: consequences for the household and broader social implications. J Nutr. 1999, 129: 525S-528S.
27. Kaiser LL, Legar-Quiñonez HR, Lamp CL, Johns MC, Sutherlin JM, Harwood JO: Food security and nutritional outcomes of preschool-age Mexican-American children. J Am Diet Assoc. 2002, 102: 924-929. 10.1016/S0002-8223(02)90210-5.
28. Rosas LG, Harley K, Fernald LCH, Guendelman S, Mejia F, Neufeld LM, Eskenazi B: Dietary associations of household food insecurity among children of Mexican descent: results of a binational study. J Am Diet Assoc. 2009, 109: 2001-2009. 10.1016/j.jada.2009.09.004.
29. Dave JM, Evans AE, Saunders RP, Watkins KW, Pfeiffer KA: Associations among food insecurity, acculturation, demographic factors, and fruit and vegetables intake at home in hispanic children. J Am Diet Assoc. 2009, 109: 697-701. 10.1016/j.jada.2008.12.017.
30. Matheson DM, Varady J, Varady A, Killen JD: Household food security and nutritional status of Hispanic children in the fifth grade. Am J Clin Nutr. 2002, 76: 210-217.
31. McIntyre L, Glanville NT, Raine KD, Dayle JB, Anderson B, Battaglia N: Do low-income lone mothers compromise their nutrition to feed their children?. CMAJ. 2003, 168 (6): 686-691.
32. Stevens CA: Exploring Food Insecurity Among Young Mothers (15-24 Years). JSPN. 2010, 15 (2): 163-171.
33. Connell CL, Nord M, Lofton KL, Yadrick K: Food security of older children can be assessed using a standardized survey instrument. J Nutr. 2004, 134: 2566-2572.
34. Nord M, Hopwood H: Recent advances provide improved tools for measuring Children's food security. J Nutr. 2007, 137: 533-536.
35. Connell CL, Lofton KL, Yadrick K, Rehner TA: Children's Experiences of food insecurity Can assist in understanding its effect on their well-being. J Nutr. 2005, 135: 1683-1690.
36. Bickel G, Nord M, Price C, Hamilton W, Cook J: Guide to Measuring Household Food Security. 2000

37. Fram MS, Frongillo EA, Jones SJ, Williams RC, Burke MP, DeLoach KP, Blake CE: Children Are aware of food insecurity and take responsibility for managing food resources. J Nutr. 2011, 141: 1114-1119. 10.3945/jn.110.135988.

38. Kaiser LL, Megar-Quiñonez H, Townsend MS, Nicholson Y, Fujii ML, Martin AC, Lamp CL: Food insecurity and food supplies in latino households with young children. J Nutr Educ Behav. 2003, 35: 148-153. 10.1016/S1499-4046(06)60199-1.

39. Division of Nutrition Physical Activity and Obesity, National Center for Chronic Disease Prevention and Health Promotion, Centers for Disease Control and Prevention: About BMI for Children and Teens. [http://www.cdc.gov/healthyweight/assessing/bmi/childrens_bmi/about_childrens_bmi.html]

40. Centers for Disease Control and Prevention (CDC)/National Center for Health Statistics: Clinical Growth Charts. [http://www.cdc.gov/growthcharts/clinical_charts]

41. Nutrition Coordinating Center: Nutrition data system for research (NDS-R) 2009. 2009, Minneapolis: Regents of the University of Minnesota

42. Sharkey JR, Branch LG, Zohoori N, Giuliani C, Busby-Whitehead J, Haines PS: Inadequate nutrient intake among homebound older persons in the community and its correlation with individual characteristics and health-related factors. Am J Clin Nutr. 2002, 76: 1435-1445.

43. U.S. Department of Agriculture and U.S. Department of Health and Human Services: Dietary Guidelines for Americans, 201. 2010, Washington: U.S. Government Printing Office, 7

44. U.S. Department of Agriculture ARS, Beltsville Human Nutrition Research Center, Food Surveys Research Group (Beltsville, MD) and U.S. Department of Health and Human Services, Centers for Disease Control and Prevention, National Center for Health Statistics (Hyattsville, MD): What We Eat in America, NHANES 2007-2008. [http://www.ars.usda.gov/Services/docs.htm?docid=18349]

45. Johnson KM, Lichter DT: Natural increase: a New source of population growth in emerging hispanic destinations in the United States. Popul Dev Rev. 2008, 34 (2): 327-346. 10.1111/j.1728-4457.2008.00222.x.

46. Grieco EM: Race and Hispanic Origin of the Foreign-Born Population in the United States: 2007. American Community Survey Reports, ACS-11. 2009, Washington, DC: U.S. Census Bureau

47. Federal Interagency Forum on Child and Family Statistics: America's Children: Key National Indicators of Well-Being. 2011, Washington, DC: U.S. Government Printing Office

48. Baxter SD, Hardin JW, Guinn CH, Royer JA, Mackelprang AJ, Smith AF: Fourth-grade Children's dietary recall accuracy is influenced by retention interval (Target period and interview time). J Am Diet Assoc. 2009, 109: 846-856. 10.1016/j.jada.2009.02.015.

49. Baxter SD, Thompson WO, Davis HC: Prompting methods affect the accuracy of children's school lunch recalls. J Am Diet Assoc. 2000, 100: 911-918. 10.1016/S0002-8223(00)00264-9.

50. Emmons L, Hayes M: Accuracy of 24-hr. recalls of young children. J Am Diet Assoc. 1973, 62: 409-415.

51. Lindquist CH, Cummings T, Goran MI: Use of tape-recorded food records in assessing Children's dietary intake. Obes Res. 2000, 8: 2-11. 10.1038/oby.2000.2.

CHAPTER 11

Obesity Prevention and National Food Security: A Food Systems Approach

Lila Finney Rutten, Amy Lazarus Yaroch, Heather Patrick, and Mary Story

11.1 INTRODUCTION

The burdens of obesity and food insecurity are unequally distributed in the USA population, with shared risk factors rendering certain socioeconomic and racial and ethnic subgroups at greater risk for both [1–3]. The intersection of obesity and food insecurity in the USA points to a public health imperative for scientists, practitioners, and policy makers to document and address food system inadequacies and leverage existing social programs to simultaneously address the nutrition issues of obesity and food insecurity [3]. Both food insecurity and obesity are increasingly recognized as forms of malnutrition resulting from poor dietary quality (higher intakes of nutrient-poor energy-dense foods) [4]. Food insecurity and obesity stem from a shared food system, therefore, corrective action must be taken within the underlying system from which they derive [5, 6].

In this paper, we briefly discuss findings from research examining associations between food insecurity and obesity in the United States (USA) and emphasize the need for greater synergy between food insecurity policies and initiatives and national public health goals around obesity prevention. We identify the common ground between these nutrition-related public health issues and call for a broadening of scope in the research and advocacy communities to align efforts around the shared goal of improving the health of at risk populations. We propose an ecological framework that identifies levers for change within the physical and social aspects of food systems to guide simultaneous attention to the pressing public health problems of food insecurity and obesity.

11.2 OBESITY AND FOOD INSECURITY

Although obesity and food insecurity historically have been viewed as distinct public health issues, there has been increasing interest in understanding the seemingly paradoxical association between obesity and food insecurity, with escalating public health concern over the dual burden of food insecurity and obesity shouldered by certain populations [3, 7, 8]. Rates of obesity have steadily increased in the USA over the past several decades, increasing dramatically among adult, child and adolescent populations [9–16]. Recent estimates from the National Health and Nutrition Examination Survey indicate that over one-third of the USA adult population is obese [12, 15]. Similar data from the National Survey of Children's Health indicate that approximately 16% of USA children are obese [17]. Obesity and related chronic disease are significant contributors to preventable morbidity and mortality and current population trends in obesity threaten to stall or reverse trends toward greater health and longevity in the USA population [9, 16, 18–20].

Food security is defined by the United States Department of Agriculture (USDA) as having access to enough food for all household members, at all times, to lead active, healthy lives [1, 21–23]. Households wherein members experience uncertainty around obtaining enough food due to insufficient resources are considered food insecure [21, 23]. Food-insecure households may experience *low food security* or *very low food security*. Households experiencing *low food security* often avoid substantially disrupting their eating patterns by engaging in coping strategies such as eating less varied diets, participating in federal food assistance programs, and accessing community feeding programs [1, 21–24]. Households experiencing *very low food security* lack sufficient resources to obtain food, which disrupts the normal eating patterns of one or more household members [1, 21–23].

Evidence from research examining the nature of the relationship between food insecurity and risk of obesity in children and adolescents is somewhat mixed [3, 7, 8, 25–37]. Recent meta-analytic reviews and research syntheses identify a relationship between food insecurity and obesity among certain subpopulations such as households with incomes below the poverty line, households led by a single adult, and households headed by African American and/or Hispanic persons [1, 38, 39]. Regardless of whether a causal association exists between food insecurity and obesity, growing evidence documents a coexistence of these nutrition-related problems [3]. Thus, collaborative efforts to prevent obesity and eradicate food insecurity in the USA are needed [3, 5, 6].

The overlap in risk factors for obesity and food insecurity is undeniable [3, 11, 32, 33, 38, 40–43]. Current evidence documents an excess burden of food insecurity and an excess burden of obesity among households living in poverty [2, 3, 40, 44]. Differential access and affordability of more nutritious food options (e.g., fruits and vegetables) have been proposed as potential contributors to existing health disparities and the higher rates of diet-related chronic disease and obesity experienced by low-income and racial ethnic minority populations [45]. The term "food desert" has been used to describe geographic areas with limited access to affordable and nutritious food. Populations at greatest risk for obesity, including those living in food deserts, are also at increased risk for experiencing food insecurity [1, 46, 47].

Efforts to address obesity and food insecurity in the USA have often been singular, yet parallel. For instance, obesity prevention efforts include the "Let's Move" campaign, as well as the push by federal and philanthropic organizations to reverse the trend of childhood obesity by 2015; concurrently there are separate initiatives to alleviate or end childhood hunger, including the pledge by President Obama to end childhood hunger by 2015. Nutrition-relevant policy changes and interventions should simultaneously attend to issues of food insecurity and obesity to ensure that efforts to control obesity do not create additional burden for those struggling with food insecurity or put more households at risk for food insecurity [3, 5]. It is equally important that efforts to reduce food insecurity do not inadvertently contribute to the obesity epidemic through provision of nutrient-poor, energy-dense foods [3].

11.3 FOOD ASSISTANCE INITIATIVES AND PUBLIC HEALTH NUTRITION GOALS

Evidence linking participation in food assistance programs and obesity has been mixed and fraught with methodological limitations [3, 48, 49]. The growing obesity epidemic in the USA, coupled with the disproportionate burden of obesity and related chronic disease shouldered by populations who access food assistance programs, represents a public health imperative to leverage existing food assistance programs and other resources to support the health and well-being of disadvantaged populations [3].

The USDA directs 15 food and nutrition programs in the USA, assisting approximately one in four Americans each year [24]. The five largest USDA administered domestic food and nutrition assistance programs include the Supplemental Nutrition Assistance Program (SNAP), the National School Lunch Program, the Special Supplemental Nutrition Program for Women, Infants, and Children (WIC), the Child and Adult Care Food Program, and the School Breakfast Program [24]. During fiscal year 2010, each of these programs expanded, proving a nutritional safety net to an increasingly food insecure population [24]. SNAP, formerly known as food stamps, was initiated in the 1960s and expanded in the 1970s to address growing rates of underconsumption and inadequate nutrient intake and is now the largest food assistance program in the USA. In 2010, SNAP served approximately 44.7 million people each month [50].

Despite increasing population reliance on federal food assistance programs, the most significant nutrition-related public health problem in the USA population today has shifted from one of deficiency to one of excess; rates of overweight and obesity and associated chronic disease have grown at an alarming rate in the USA population over the past three decades, particularly among low-income ethnic minority populations [48]. Existing evidence documents an excess burden of obesity among populations with lower socioeconomic position and among food insecure populations [2, 38–40, 44]. Interest in identifying policy and programmatic actions to address this public health disparity is mounting [5].

The disproportionately higher incidence of overweight and obesity and associated chronic disease among certain low-income populations has stimulated public health researchers and policy makers to explore ways in which food assistance programs, such as SNAP, might be modified to improve the dietary quality of recipients and thereby prevent or reduce obesity [49]. Proposals to facilitate and/or place restrictions on the types of foods that can be purchased with food supplement benefits (e.g., increased fruits and vegetables and decreased sugar sweetened beverages), to provide incentives for buying more healthful foods, and to expand nutrition education efforts have been suggested as means to support improvements in dietary quality among recipients to reduce risk factors for obesity and related chronic disease [51]. Improvements in federal food assistance programs, including the income supplement for the purchase of fruits and vegetables in the SNAP Healthy Incentive Program, Electronic Benefit Transfer (EBT), and Double Up Food Bucks Programs to promote purchase and consumption of local produce at farmers markets, modification of the WIC food package to include vouchers for fruits and vegetables and changes to meals in the National School Lunch Program to adhere to USDA nutrition guidelines, heed the call for greater synergy in efforts to reduce obesity and improve food security [27].

11.4   PUBLIC HEALTH GOALS AND RESEARCH PRIORITIES AROUND OBESITY AND FOOD INSECURITY

Current public health goals around obesity prevention are beginning to reflect an understanding of the power of public policy and environmental change to influence both individual- and population-level behavioral change while acknowledging the importance of promoting population food security as part of such efforts. Implementation of public policy at a macrolevel, including local, state, or federal legislation, is an effective and comprehensive means to affect change in population behavior. Policy-level changes influence the environments that subsequently influence the behavior and choices of individuals. Effective public health policy and environmental change can encourage populations to practice more healthful behavior [52]. That is, effective policies can change environments to have improved access (and lower costs) to healthier foods and to subsequently make the healthy choice the easy or default choice [52–57].

The *Nutrition and Weight Status* objectives identified for Healthy People 2020 are centered on an overarching goal to promote health and reduce chronic disease risk through healthful diets and maintenance of healthy body weights [58]. The stated objectives emphasize that efforts to improve diet and reduce obesity require attention to the policies and environments that support individual behavior across a variety of settings. Increasing household food security and eliminating food insecurity are specified in the Healthy People 2020 objectives as integral to goals of promoting healthful diets and healthy weight [58].

Responding to the growing childhood obesity epidemic, the White House Task Force on Childhood Obesity developed and is working to implement an interagency plan to eliminate child obesity [59]. The Task Force published an action plan with specific recommendations for addressing the childhood obesity epidemic, including a call for further research and related policy to address food insecurity [59]. During the first year of this initiative, the Healthy, Hunger-Free

Kids Act was enacted to expand children's access to healthy school meals [60]. This act aims to increase the number of eligible children enrolled in school meal programs through direct certification of children who receive other federal assistance and improve access to nutritious meals in schools [60].

The National Institutes of Health (NIH) has also convened an Obesity Research Task Force to develop a strategic plan to accelerate multidisciplinary research efforts to address the population burden of obesity and related chronic disease. The task force identified and recently published a series of research challenges and opportunities to inform NIH research planning [61]. Importantly, the NIH Strategic Plan for Obesity Research explicates the need to integrate research results into community programs and clinical practice [61]. Historically, this need has not plainly taken into account food insecurity. However, given the congruency in population health goals around chronic disease prevention and overlap in risk factors and mechanisms, it is apparent that efforts to promote food security should be woven more closely with obesity prevention.

11.5   A FOOD SYSTEMS APPROACH TO REDUCING FOOD INSECURITY AND PREVENTING OBESITY

Public health advocates have long sought to unify food assistance efforts with high-quality nutrition. Focused attention on how current food system policies and practices may impact public health and diet-related chronic disease and obesity is essential to developing a systematic strategy that simultaneously achieves community food justice and public health success [5, 6]. The American Public Health Association has defined a sustainable food system as "one that provides healthy food to meet current food needs while maintaining healthy ecosystems that can also provide food for generations to come with minimal negative impact to the environment. A sustainable food system also encourages local production and distribution infrastructures and makes nutritious food available, accessible, and affordable to all." [62]. Furthermore, per the Healthy People 2020 Nutrition and Weight Status objectives, greater attention to food insecurity, food systems, and food justice is encouraged in public health efforts to reduce or prevent obesity [58].

Our call for a systematic approach to simultaneously address the pressing public health issues of food insecurity and obesity points to an opportunity that is ripe for the integration of the existing evidence and further strategic development of the science around food insecurity and obesity to inform public policy and community programs about food with concurrent consideration of both issues [5, 6]. Efforts to close the gap between science and public policy around food security and obesity will require changes in both the research and advocacy communities to align efforts around the shared goal of improving the health and well-being of at-risk populations [5, 6]. Strategic science aimed at changing public understanding of the link between food insecurity and obesity coupled with science-based efforts to inform agricultural and nutrition legislation, policy and regulations offers a useful paradigm for harnessing obesity prevention and nutrition science to change public policy with relevance to food security.

Typical approaches to public health problems with behavioral or lifestyle components, such as obesity and food insecurity, have focused mainly on individual change through motivational or educational interventions. This approach, as applied at the intersection of food insecurity and obesity, has been a failed experiment. Despite decades of individual-focused interventions for obesity and expansion of investments in nutrition education for food assistance recipient [58], we have not reversed the trend of either condition. To effectively reduce the public health burden of obesity in food insecure populations, science and advocacy efforts must align focus on systemic and environmental factors to cultivate "optimal defaults" for individual nutrition-related behavior by making the healthful option the easy option. {Thaler, 2008 number 51}. An example of an optimal default may be placing fruits and vegetables at the front of school cafeteria lines to make them more accessible or including apple slices as the "default" side dish instead of French fries in children's meals.

The processes underlying food insecurity and obesity derive from a shared food system [5]. Therefore, efforts to address food insecurity and obesity must emulate the complexity of the system from which they emerge. The food systems concept describes the required inputs, processes, and generated outputs involved in the provision of food and nutrients for sustenance and health including growing, harvesting, processing, packaging, transporting, marketing, consuming, and disposing of food [63, 64]. A set of shared food system principles supporting economically, ecologically, and socially sustainable food systems that promote the health of individuals, communities, and the environment have been developed through the cooperative efforts of the Academy of Nutrition and Dietetics (formerly the American Dietetic Association), the American Nurses Association, the American Planning Association, and the American Public Health Association [65]. Accordingly, health-promoting food systems are defined as those that support the physical and mental health of producers and consumers with accountability for the public health impact across the food system including production, processing, packaging, labeling, distribution, consumption, and disposal. Sustainability of food systems thus derives from the conservation and regeneration of natural resources and biodiversity to manage current food and nutrition requirements with transparency and equity in process and outcomes without compromising future system efficacy and output [63, 64]. Health-promoting and sustainable food systems are necessarily diverse in size and scale, geography, and culture to ensure resiliency in the face of ecological and economic challenges and to promote diversity and equity in the availability of healthful food options [63, 64].

Adopting a food systems approach facilitates awareness of the complexity of food systems and of the social, economic, environmental, and political contexts within which they operate [63, 64]. Aligning food assistance initiatives with public health nutrition goals requires a systematic approach to simultaneously address food insecurity and obesity through research and practice efforts aimed at establishing sustainable food systems to promote health, improve food security, and prevent obesity [5]. Individual, environmental, and policy level changes to promote the development of healthy sustainable food systems that simultaneously address the nutrition-related public health goals of obesity prevention and food security are needed.

11.6 LEVERS FOR CHANGE IN THE MACRO- AND MICRO-FOOD SYSTEM ENVIRONMENTS

Several opportunities exist for cultivating greater synergy between public health efforts around food insecurity and obesity prevention. Delineation of areas for potential intervention and public health impact can be organized according to the ecological perspective inherent in the food systems approach. Specifically, priorities for addressing shared risk factors and health outcomes in obesity prevention and food security can be identified and operationalized according to the size and nature of the environments in which intervention is to be implemented. This approach, originally described as the analysis grid for environments linked to obesity, conceptualizes environments in terms of size and type to identify potential interventions [66].

This ecological approach can be adapted and expanded to conceptualize the shared physical, economic, political, and sociocultural environments of obesity and food insecurity and to identify levers for change and opportunities for intervention that may impact individual behavior (see Table 1). Within this

TABLE 11.1 Levers for change in the macro- and microenvironments to influence obesity in food insecure populations.

	Environment type			
	Physical	Economic	Political	Sociocultural
Macroenvironmental settings	Modify agricultural, housing, transportation, and social policies that influence food production and distribution.	Offer monetary incentives for healthy food options (e.g., subsidies) and disincentives for unhealthy options (e.g., taxes).	Promote agricultural, social, and food security and nutrition policy informed by obesity prevention science.	Food marketing and advertising environment in economically disadvantaged areas to promote health and prevent obesity.
Microenvironmental setting	Food retailers and food service outlets determine local healthy food options.	Institutional financial support for health promotion and nutrition programs; financial support for support local food production.	Institutional rules and policies influencing availability of healthy food options.	Institutional climate around nutritious eating and healthy body weight maintenance.

framework, macroenvironmental sectors describe the greater influencing context of industry and supporting infrastructure on available and consumed food options [66]. For example, macroenvironmental sectors relevant to the food environment may include food production and manufacturing, distribution, and marketing as well as relevant technological and social infrastructure. The framework describes microenvironmental settings as the settings that involve food where people gather for specific purposes [66]. Examples of microenvironments include homes, schools, workplaces, community venues, food service or retail outlets (e.g., supermarkets, restaurants, etc.), and healthcare settings. Within both macroenvironmental sectors and microenvironmental settings, several environmental types with relevance to obesity prevention and food insecurity have been identified including the physical environment, the economic environment, the political environment, and the sociocultural environment [66]. We examine each of these types, in turn, to identify effective interventions to concurrently address obesity and food insecurity.

11.6.1   Physical Food Environment

The physical food environment refers to the availability of food outlets, relevant training opportunities, and nutrition-related information within specific settings (e.g., supermarkets, communities, schools, etc.) [66]. Identifying levers for change within the physical environment requires understanding the ways in which macroenvironmental physical factors including resource inputs, food production, distribution, and transportation systems influence the availability of healthy food venues by geographic region. Resource inputs refer to the raw materials, biophysical factors, and social factors available in a given environment for input into the food system. Food production involves transformation of resource inputs into raw agricultural goods, and food processing involves transformation of production output agricultural goods into food for distribution. Many low-income populations live in "food deserts" wherein failures in distribution result in geographic areas devoid of retail establishments offering healthy food options thereby limiting access to affordable, nutritious food in such neighborhoods [67, 68].

Opportunities for intervention in the macroenvironmental physical food environment may legislative efforts to modify existing agricultural, housing, transportation, and social policies to address the dual issues of food security and obesity prevention [5, 68]. For example, currently retailers authorized to participate in SNAP are required to sell staple foods for home preparation and consumption and must offer on a continuous basis, at least three different varieties of foods from the four staple food categories, with at least two of the food categories being perishable foods [68]. Alternatively, authorized SNAP retailers must have more than half of their total gross sales from staple foods. This policy aims to increase access to healthy food options, although most SNAP retailers are authorized under the first criterion and therefore can meet said requirements by offering a small number and variety of staple food items [68]. Efforts are underway by the USDA to review current regulations with the goal of striking a balance to maintain an adequate supply of required foods while retaining retailer participation in SNAP [68]. Modifications to WIC food packages to include whole grains, fruits and vegetables, and reduced fat milk (versus whole milk) for children over 2 years of age have also been successful in improving the diets of WIC recipients [54, 57, 69]. Other policy level interventions at the macroenvironmental level that may influence the physical food environment include incorporation of community food access into housing and community development planning and transportation development and planning and related policy [67, 68].

Food distribution determines the local availability of points of access including wholesale or retail entities, the food service industry, and public and private food assistance programs and is therefore the vital link to consumer acquisition at the microenvironmental setting level. At the microenvironmental level, grassroots and local community food projects to improve availability of food options have shown promise in improving access and the dietary intake in at risk populations [68, 70]. The USDA Community Food Projects Competitive Grants Program has funded hundreds of community food projects under the umbrella aim of supporting communities in local efforts to improve local food systems [70]. Examples of community food project activities may include: increasing the availability of locally produced, healthy food options through community gardens, farmers markets, and food assistance programs; improving dietary composition through nutrition education, cooking classes, and engagement in food production; increasing participation in nutrition programs; and working to integrate food system issues into community planning and local public health initiatives [68, 70]. Although there is considerable diversity in community food projects as per the needs of diverse communities, a systematic review of five years of funded projects revealed the following shared aims of community food projects: they focus on the food needs of low-income populations; they aim to connect local food producers and consumers; they strive to increase the local food production and self-reliance; and they attempt to develop integrated solutions to agriculture, food, and nutrition-related problems [70]. These goals coincide with community food security and obesity prevention goals and strive to develop local food systems that promote health, sustainability, and community self-sufficiency.

Opportunities for intervention may also occur at the point of purchase, wherein the concept of optimal defaults could be applied to location of healthy food options in retail and wholesale settings to promote consumption [52]. Possibilities for dissemination of nutrition information also abound, with product and menu labeling, and in-store educational campaigns aimed at raising consumer awareness of healthy and affordable food options [71–73]. Addressing physical environment food system vulnerabilities requires identification of specific deficits in human and/or technological resources resulting in food system failure at the point of resource inputs, production, processing, distribution, or point of purchase (e.g., food labeling) that must be addressed in order to produce and distribute adequate and healthful food resources for a given population [63].

11.6.2   Economic Food Environment

The economic environment describes the costs related to food production, manufacturing, distribution, retailing, and purchase [66]. Economic factors at both the macro- and microenvironmental level are strongly related to both obesity and food insecurity [2, 38, 40, 44]. Expansion of federal antipoverty initiatives, including food and nutrition assistance programs such as SNAP and WIC, school meal programs, the Head Start program, the Earned Income Tax Credit, and the Temporary Assistance for Needy Families program, may serve to improve access to healthy food options [48, 74–76].

Individuals may also lack the necessary resources, awareness, and/or skills to obtain consistent access to nutritious food. Monetary incentives, such as food subsidies, for production and consumption of healthy food options and monetary disincentives for the production and consumption of unhealthy food options, such as taxes, represent potentially powerful levers for change in both obesity prevention and food insecurity [66]. Other promising interventions to prevent obesity and reduce food insecurity may include institutional financial support for health promotion and nutrition programs and financial support to promote local food production efforts [68]. Efforts to incorporate nutrition education into existing programs, including school-based programs, Head Start, WIC, and SNAP, that reach at risk populations have been successful in improving nutrition-related knowledge and behavior and are particularly encouraged [75–79].

11.6.3   Political Food Environment

The political food environment refers to the laws, regulations (macroenvironment), and institutional rules (microenvironment) that influence available food options and related individual behavior [66]. The political environment holds prominent opportunities for wide-spread change in population burden of food insecurity and obesity. As per our prior discussion of policy level interventions to influence the physical and economic food environments, agricultural, transportation, social legislation, and resultant food and nutrition policy informed by obesity prevention science and public health goals around food security could promote the health and well-being of the population [45]. At the macrolevel, additional regulatory forces in the political environment determine nutrition labeling standards, health claims on packages and in stores, and the nature of food advertisements aimed at youth. Packaging, labeling, and advertisement exert considerable influence on consumer behavior, and therefore represent significant levers for change in the political environment. At the microlevel, schools, hospitals, and workplaces may adopt policies regarding the nature of food services, including requirements for the quantities and qualities of foods served in cafeterias, vending machines, and other outlets [80, 81].

11.6.4   Sociocultural Food Environment

The sociocultural food environment refers to the social and cultural norms or beliefs, values, and attitudes about food, embraced by a community or society [66]. At the macroenvironmental level, food marketing and advertising can be leveraged as compelling interventions to promote health and prevent obesity. At the microenvironmental level, institutional climate around nutritious eating and healthy body weight maintenance can be targeted within specific settings (e.g., schools or worksites) to encourage and educate around healthful and affordable food options. Homes also represent an important microenvironmental context in which social and cultural norms about food are shaped. The home environment and parenting styles and practices, in particular, have received increasing attention with regard to the role that parents play in children's eating behaviors. For example, parents are not only responsible for the amounts and types of food made available in the home; they are the first socializing agent for children's eating. A large body of research has demonstrated that children's eating is shaped instrumentally by the foods that parents bring into the home and by modeling. A growing evidence base suggests that the ways in which parents go about attempting to shape children's eating is also important. For example, a parent may limit a child's junk food intake by restrictive practices such as not allowing certain foods in the home at all or vigilantly monitoring their child's eating behaviors. By comparison, a parent may also limit junk food intake by talking to the child about why salty and sugary snack foods are "sometimes" foods to be eaten only in small amounts or on special occasions. Evidence suggests that parental feeding practices that include behaviors such as providing age-appropriate rationales and some structure and limit settings (i.e., a middle ground between allowing children to make all of the decisions and forbidding eating some foods like junk food while demanding eating others like fruit and vegetables) result in children eating and self-regulating food intake better [82–84].

Although there is evidence that parental feeding styles and practices influence children's eating behaviors, less is known about the precursors to parental feeding styles and practices. Food insecurity may be one such factor. Indeed, when parents are food insecure it is likely that they will interact with their children around eating differently than if they are confident that they will have sufficient food for the next meal, the next day, or for the next month. For example, they may be more likely to encourage children to eat more when it is available, even after the child has indicated that he or she is full. Literature on the parental feeding practice of "clean your plate" (i.e., eat everything that is on your plate) indicates that this is one mechanism by which children stop paying attention to physiological cues of hunger and fullness and instead rely on cues in the physical environment [85]. More recent evidence suggests that pressuring children to eat may backfire in terms of children's food preferences, which has important implications not only for the quantity of foods consumed but for their quality [86]. Although food insecure parents may be engaging in this practice for good reason, the potential long-term consequences of encouraging children to eat past the physiological sense of fullness have important implications for eating regulation and obesity. Food insecure parents may also be more likely to purchase and serve foods that they are certain their children will eat (e.g., sugary or salty food), because they do not want to risk-wasting resources on foods that their children may not eat.

11.7 CONCLUSION

Increasing population burden of food insecurity and obesity speaks to the critical need for development of a comprehensive approach to reform existing food systems to simultaneously address issues of community food justice, food security, food quality, and public health success related to obesity prevention. Interventions that cultivate sustainable food systems to promote health, improve food security, and prevent obesity, with multiple social, ecological, and economic benefits, based on an ecological approach that conceptualizes the shared physical, economic, political, and sociocultural environments of obesity and food insecurity are needed. Current food insecurity initiatives and national obesity prevention public health goals could be coordinated through the adoption of a food systems approach to conduct strategic science to inform public health interventions aimed at improving population health through environmental and policy level changes to promote nutrition environments with optimal defaults to support individual behavior. Continued efforts are needed to systematically identify policy gaps and opportunities for and barriers to merging food security and obesity prevention initiatives, as part of an ongoing process of developing and implementing an integrated and comprehensive strategy for addressing the nutrition-related needs of at risk populations.

REFERENCES

1. M. Nord, M. Andrews, and S. Carlson, Household Food Security in the United States, 2008, U.S. Department of Agriculture, Economic Research Service, Washington, DC, USA, 2009.
2. K. Ball and D. Crawford, "Socio-economic factors in obesity: a case of slim chance in a fat world?" Asia Pacific Journal of Clinical Nutrition, vol. 15, supplement, pp. 15–20, 2006.
3. N. I. Larson and M. T. Story, "Food insecurity and weight status among U.S. children and families: a review of the literature," American Journal of Preventive Medicine, vol. 40, no. 2, pp. 166–173, 2011.
4. A. L. Yaroch and C. A. Pinard, "Are the hungry more at risk for eating calorie-dense nutrient-poor foods?: comment on "first foods most: after 18-hour fast, people drawn to starches first and vegetables last" the hungry eat calorie-dense nutrient-poor foods," Archives of Internal Medicine, vol. 172, no. 12, pp. 963–964, 2012.
5. L. Dube, P. Pingali, and P. Webb, "Paths of convergence for agriculture, health, and wealth," Proceedings of the National Academy of Sciences, vol. 109, no. 31, pp. 12294–12301, 2012.
6. R. A. Hammond and L. Dubé, "A systems science perspective and transdisciplinary models for food and nutrition security," Proceedings of the National Academy of Sciences, vol. 109, no. 31, pp. 12356–12363, 2012.
7. J. Bhattacharya, J. Currie, and S. Haider, "Poverty, food insecurity, and nutritional outcomes in children and adults," Journal of Health Economics, vol. 23, no. 4, pp. 839–862, 2004.

8. L. M. Dinour, D. Bergen, and M. C. Yeh, "The food insecurity-obesity paradox: a review of the literature and the role food stamps may play," Journal of the American Dietetic Association, vol. 107, no. 11, pp. 1952–1961, 2007.

9. NIH, "Clinical guidelines on the identification, evaluation, and Treatment of overweight and obesity in adults—the evidence report. National institutes of health," Obesity Research, vol. 6, supplement 2, pp. 51S–209S, 1998.

10. CDC, "State-specific prevalence of obesity among adults—United States, 2007," Morbidity and Mortality Weekly Report, vol. 57, no. 28, pp. 765–768, 2008.

11. A. H. Mokdad, M. K. Serdula, W. H. Dietz, B. A. Bowman, J. S. Marks, and J. P. Koplan, "The spread of the obesity epidemic in the United States, 1991–1998," Journal of the American Medical Association, vol. 282, no. 16, pp. 1519–1522, 1999.

12. K. M. Flegal, M. D. Carroll, C. L. Ogden, and L. R. Curtin, "Prevalence and trends in obesity among US adults, 1999–2008," Journal of the American Medical Association, vol. 303, no. 3, pp. 235–241, 2010.

13. K. M. Flegal, M. D. Carroll, C. L. Ogden, and L. R. Curtin, "Prevalence and trends in obesity among US adults, 1999–2008," Journal of the American Medical Association, vol. 303, no. 3, pp. 235–241, 2010.

14. C. L. Ogden, M. D. Carroll, L. R. Curtin, M. M. Lamb, and K. M. Flegal, "Prevalence of high body mass index in US children and adolescents, 2007-2008," Journal of the American Medical Association, vol. 303, no. 3, pp. 242–249, 2010.

15. M. Shields, M. D. Carroll, and C. L. Ogden, "Adult obesity prevalence in Canada and the United States," Tech. Rep. 56, National Center for Health Statistics Data Brief, 2011, http://www.cdc.gov/nchs/data/databriefs/db56.htm.

16. S. J. Olshansky, D. J. Passaro, R. C. Hershow et al., "A potential decline in life expectancy in the United States in the 21st century," The New England Journal of Medicine, vol. 352, no. 11, pp. 1138–1145, 2005.

17. G. K. Singh, M. D. Kogan, and P. C. Van Dyck, "Changes in state-specific childhood obesity and overweight prevalence in the United States from 2003 to 2007," Archives of Pediatrics and Adolescent Medicine, vol. 164, no. 7, pp. 598–607, 2010.

18. M. H. Park, C. Falconer, R. M. Viner, and S. Kinra, "The impact of childhood obesity on morbidity and mortality in adulthood: a systematic review," Obesity Reviews, vol. 13, no. 11, pp. 985–1000, 2012.

19. J. J. Reilly and J. Kelly, "Long-term impact of overweight and obesity in childhood and adolescence on morbidity and premature mortality in adulthood: systematic review," International Journal of Obesity, vol. 35, no. 7, pp. 891–898, 2011.

20. M. Lenz, T. Richter, and I. Mühlhauser, "The morbidity and mortality associated with overweight and obesity in adulthood: a systematic review," Deutsches Arzteblatt, vol. 106, no. 40, pp. 641–648, 2009.

21. M. Nord, M. Andrews, and S. Carlson, Household Food Security in the United States, 2007, U.S. Department of Agriculture, Economic Research Service, Washington, DC, USA, 2008.

22. M. Nord, A. Coleman-Jensen, M. Andrews, and S. Carlson, Household Food Security in the United States, 2009, US Department of Agriculture, Economic Research Service, Washington, DC, USA, 2010.

23. M. Nord, M. Andrews, and S. Carlson, Household Food Security in the United States, 2010, US Department of Agriculture, Economic Research Service, Washington, DC, USA, 2011.

24. ERS, "The food assistance landscape: fiscal year 2010 annual report," Economic Information Bulletin 6–8, Economic Research Service, U.S. Department of Agriculture, Washington, DC, USA, 2011.

25. P. H. Casey, P. M. Simpson, J. M. Gossett et al., "The association of child and household food insecurity with childhood overweight status," Pediatrics, vol. 118, no. 5, pp. e1406–e1413, 2006.

26. P. H. Casey, K. Szeto, S. Lensing, M. Bogle, and J. Weber, "Children in food-insufficient, low-income families: prevalence, health, and nutrition status," Archives of Pediatrics and Adolescent Medicine, vol. 155, no. 4, pp. 508–514, 2001.

27. P. B. Crawford and K. L. Webb, "Unraveling the paradox of concurrent food insecurity and obesity," American Journal of Preventive Medicine, vol. 40, no. 2, pp. 274–275, 2011.

28. S. J. Jones and E. A. Frongillo, "Food insecurity and subsequent weight gain in women," Public Health Nutrition, vol. 10, no. 2, pp. 145–151, 2007.

29. D. F. Jyoti, E. A. Frongillo, and S. J. Jones, "Food insecurity affects school children's academic performance, weight gain, and social skills," Journal of Nutrition, vol. 135, no. 12, pp. 2831–2839, 2005. A. Karnik, B. A. Foster, V. Mayer et al., "Food insecurity and obesity in New York City primary care clinics," Medical Care, vol. 49, no. 7, pp. 658–661, 2011.

30. K. S. Martin and A. M. Ferris, "Food insecurity and gender are risk factors for obesity," Journal of Nutrition Education and Behavior, vol. 39, no. 1, pp. 31–36, 2007.

31. E. Metallinos-Katsaras, B. Sherry, and J. Kallio, "Food insecurity is associated with overweight in children younger than 5 years of age," Journal of the American Dietetic Association, vol. 109, no. 10, pp. 1790–1794, 2009.

32. A. F. Meyers, R. J. Karp, and J. G. Kral, "Poverty, food insecurity, and obesity in children," Pediatrics, vol. 118, no. 5, pp. 2265–2266, 2006.

33. A. M. Olson and M. S. Strawderman, "The relationship between food insecurity and obesity in rural childbearing women," Journal of Rural Health, vol. 24, no. 1, pp. 60–66, 2008.

34. J. Stuff, et al., "Household food insecurity and obesity, chronic disease, and chronic disease risk factors," Journal of Hunger & Environmental Nutrition, vol. 1, no. 2, pp. 43–61, 2006.

35. R. C. Whitaker and A. Sarin, "Change in food security status and change in weight are not associated in urban women with preschool children," Journal of Nutrition, vol. 137, no. 9, pp. 2134–2139, 2007.

36. P. E. Wilde and J. N. Peterman, "Individual weight change is associated with household food security status," Journal of Nutrition, vol. 136, no. 5, pp. 1395–1400, 2006.

37. V. Shrewsbury and J. Wardle, "Socioeconomic status and adiposity in childhood: a systematic review of cross-sectional studies 1990–2005," Obesity, vol. 16, no. 2, pp. 275–284, 2008.

38. Y. Wang and M. A. Beydoun, "The obesity epidemic in the United States—gender, age, socioeconomic, racial/ethnic, and geographic characteristics: a systematic review and meta-regression analysis," Epidemiologic Reviews, vol. 29, no. 1, pp. 6–28, 2007.

39. P. M. Lantz, J. S. House, J. M. Lepkowski, D. R. Williams, R. P. Mero, and J. Chen, "Socioeconomic factors, health behaviors, and mortality: results from a nationally representative prospective study of US adults," Journal of the American Medical Association, vol. 279, no. 21, pp. 1703–1708, 1998.

40. B. A. Laraia, A. M. Siega-Riz, C. Gundersen, and N. Dole, "Psychosocial factors and socioeconomic indicators are associated with household food insecurity among pregnant women," Journal of Nutrition, vol. 136, no. 1, pp. 177–182, 2006.

41. G. K. Singh, M. D. Kogan, and S. M. Yu, "Disparities in obesity and overweight preva-
    lence among us immigrant children and adolescents by generational status," Journal of
    Community Health, vol. 34, no. 4, pp. 271–281, 2009.

42. G. K. Singh, M. Siahpush, and M. D. Kogan, "Rising Social Inequalities in US childhood
    obesity, 2003–2007," Annals of Epidemiology, vol. 20, no. 1, pp. 40–52, 2010.

43. G. Enzi, "Socioeconomic consequences of obesity: the effect of obesity on the individual,"
    PharmacoEconomics, vol. 5, supplement 1, pp. 54–57, 1994.

44. M. Story, M. W. Hamm, and D. Wallinga, "Food systems and public health: linkages to
    achieve healthier diets and healthier communities," Journal of Hunger and Environmental
    Nutrition, vol. 4, no. 3-4, pp. 219–224, 2009.

45. L. J. F. Rutten, A. L. Yaroch, U. Colón-Ramos, W. Johnson-Askew, and M. Story, "Poverty,
    food insecurity, and obesity: a conceptual framework for research, practice, and policy,"
    Journal of Hunger and Environmental Nutrition, vol. 5, no. 4, pp. 403–415, 2010.

46. M. Kursmark and M. Weitzman, "Recent findings concerning childhood food insecurity,"
    Current Opinion in Clinical Nutrition and Metabolic Care, vol. 12, no. 3, pp. 310–316, 2009.

47. M. V. Ploeg, L. Mancino, B. H. Lin, and J. Guthrie, "US food assistance programs and trends
    in children's weight," International Journal of Pediatric Obesity, vol. 3, no. 1, pp. 22–30,
    2008.

48. E. A. Frongillo, "Understanding obesity and program participation in the context of poverty
    and food insecurity," Journal of Nutrition, vol. 133, no. 7, pp. 2117–2118, 2003.

49. USDA, The food Assistance Landscape: FY 2011 Annual Report, USDA, Washington, DC,
    USA, 2012.

50. J. F. Guthrie, E. Frazao, M. Andrews, and D. Smallwood, Improving Food Choices—Can
    Food Stamps Do More? Perspectives on Food and Farm Policy: Food and Nutrition in
    Amberwaves, U.S. Department of Agriculture, Economic Research Service, Washington,
    DC, USA, 2007.

51. R. H. Thaler and C. R. Sunstein, Nudge: Improving Decisions on Health, Wealth, and
    Happiness, Yale University Press, New Haven, Conn, USA, 2008.

52. A. S. Hanks, D. R. Just, L. E. Smith, and B. Wansink, "Healthy convenience: nudging stu-
    dents toward healthier choices in the lunchroom," Journal of Public Health, vol. 34, no. 3,
    pp. 370–376, 2012.

53. Hillier, J. McLaughlin, C. C. Cannuscio, M. Chilton, S. Krasny, and A. Karpyn, "The impact
    of WIC food package changes on access to healthful food in 2 low-income urban neighbor-
    hoods," Journal of Nutrition Education and Behavior, vol. 44, no. 3, pp. 210–216, 2012.

54. J. E. Painter, B. Wansink, and J. B. Hieggelke, "How visibility and convenience influence
    candy consumption," Appetite, vol. 38, no. 3, pp. 237–238, 2002.

55. Wansink, "Environmental factors that increase the food intake and consumption volume of
    unknowing consumers," Annual Review of Nutrition, vol. 24, pp. 455–479, 2004.

56. S. E. Whaley, L. D. Ritchie, P. Spector, and J. Gomez, "Revised WIC food package improves
    diets of WIC families," Journal of Nutrition Education and Behavior, vol. 44, no. 3, pp. 204–
    209, 2012.

57. USDHHS, Healthy People 2020 Nutrition and Weight Status Objectives, Office of Disease
    Prevention and Health Promotion, U.S. Department of Health and Human Services,
    Washington, DC, USA, 2010.

58. WHTF, Report to the President: Solving the Problem of Childhood Obesity within a
    Generation, White House Task Force on Childhood Obesity (WHTF), 2010.

59. WHTF, Report to the President: Solving the Problem of Childhood Obesity within a Generation, One Year Progress Report, White House Task Force on Childhood Obesity (WHTF), 2011.

60. USDHHS, Strategic Plan for NIH Obesity Research: A Report of the NIH Obesity Research Task Force, U.S. Department of Health and Human Services, National Institutes of Health, Washington, DC, USA, 2011.

61. APHA, "Toward a healthy, sustainable food system," in American Public Health Association Policy Statement Database, American Public Health Association, 2007.

62. ADA AMA. APA, A., Principles of a healthy, sustainable food system, American Dietetic Association, the American Nurses Association, The American Planning Association, and the American Public Health Association, 2010.

63. Hawkes, "Identifying innovative interventions to promote healthy eating using consumption-oriented food supply chain analysis," Journal of Hunger and Environmental Nutrition, vol. 4, no. 3-4, pp. 336–356, 2009.

64. G. Egger, S. Pearson, S. Pal, and B. Swinburn, "Dissecting obesogenic behaviours: the development and application of a test battery for targeting prescription for weight loss," Obesity Reviews, vol. 8, no. 6, pp. 481–486, 2007.

65. T. Giang, A. Karpyn, H. B. Laurison, A. Hillier, and R. D. Perry, "Closing the grocery gap in underserved communities: the creation of the Pennsylvania fresh food financing initiative," Journal of Public Health Management and Practice, vol. 14, no. 3, pp. 272–279, 2008.

66. J. Beaulac, E. Kristjansson, and S. Cummins, "A systematic review of food deserts, 1966–2007," Preventing Chronic Disease, vol. 6, no. 3, p. A105, 2009. View at Google Scholar · View at Scopus

67. M. Ver Ploeg, V. Breneman, T. Farrigan, et al., "Access to affordable and nutritious food—measuring and understanding food deserts and their consequences: report to congress," in Administrative Publication, USDA, Economic Research Service, 2009.

68. WIC, Revisions in the WIC Food Packages, Special Supplemental Nutrition Program for Women, Infants and Children (WIC), 2007.

69. K. Pothukuchi, Building Community Food Security: Lessons from Community Food Projects, 1999–2003, 2007.

70. G. X. Ayala, M. N. Laska, S. N. Zenk, et al., "Stocking characteristics and perceived increases in sales among small food store managers/owners associated with the introduction of new food products approved by the special supplemental nutrition program for women, infants, and children," Public Health Nutrition, vol. 14, pp. 1–9, 2012.

71. N. C. Crespo, J. P. Elder, G. X. Ayala, et al., "Results of a multi-level intervention to prevent and control childhood obesity among latino children: the aventuras para ninos study," Annals of Behavioral Medicine, vol. 43, no. 1, pp. 84–100, 2012.

72. J. Gittelsohn, M, N. Laska, T. Andreyeva, et al., "Small retailer perspectives of the 2009 women, infants and children program food package changes," American Journal of Health Behavior, vol. 36, no. 5, pp. 655–665, 2012.

73. R. A. Gooze, C. C. Hughes, D. M. Finkelstein, and R. C. Whitaker, "Reaching staff, parents, and community partners to prevent childhood obesity in head start, 2008," Preventing Chronic Disease, vol. 7, no. 3, p. A54, 2010.

74. C. Hughes, R. A. Gooze, D. M. Finkelstein, and R. C. Whitaker, "Barriers to obesity prevention in head start," Health Affairs, vol. 29, no. 3, pp. 454–462, 2010.

75. R. C. Whitaker, R. A. Gooze, C. C. Hughes, and D. M. Finkelstein, "A national survey of obesity prevention practices in head start," Archives of Pediatrics and Adolescent Medicine, vol. 163, no. 12, pp. 1144–1150, 2009.

76. R. A. Gooze, C. C. Hughes, D. M. Finkelstein, and R. C. Whitaker, "Reaching staff, parents, and community partners to prevent childhood obesity in head start, 2008," Preventing chronic disease, vol. 7, no. 3, p. A54, 2010.

77. E. Wall, C. Least, J. Gromis, and B. Lohse, "Nutrition education intervention improves vegetable-related attitude, self-efficacy, preference, and knowledge of fourth-grade students," Journal of School Health, vol. 82, no. 1, pp. 37–43, 2012.

78. B. MkNelly, S. Nishio, C. Peshek, and M. Oppen, "Community health centers: a promising venue for supplemental nutrition assistance program education in the central valley," Journal of Nutrition Education and Behavior, vol. 43, no. 4, supplement 2, pp. S137–S144, 2011.

79. M. S. Nanney, T. Nelson, M. Wall et al., "State school nutrition and physical activity policy environments and youth obesity," American Journal of Preventive Medicine, vol. 38, no. 1, pp. 9–16, 2010.

80. M. Story, M. S. Nanney, and M. B. Schwartz, "Schools and obesity prevention: creating school environments and policies to promote healthy eating and physical activity," Milbank Quarterly, vol. 87, no. 1, pp. 71–100, 2009.

81. A. K. Ventura and L. L. Birch, "Does parenting affect children's eating and weight status?" International Journal of Behavioral Nutrition and Physical Activity, vol. 5, article 15, 2008.

82. H. R. Clark, E. Goyder, P. Bissell, L. Blank, and J. Peters, "How do parents' child-feeding behaviours influence child weight? Implications for childhood obesity policy," Journal of Public Health, vol. 29, no. 2, pp. 132–141, 2007.

83. J. S. Savage, J. O. Fisher, and L. L. Birch, "Parental influence on eating behavior: conception to adolescence," Journal of Law, Medicine and Ethics, vol. 35, no. 1, pp. 22–34, 2007.

84. L. L. Birch and J. O. Fisher, "Development of eating behaviors among children and adolescents," Pediatrics, vol. 101, no. 3, pp. 539–549, 1998.

85. A. T. Galloway, L. M. Fiorito, L. A. Francis, and L. L. Birch, "'Finish your soup': counterproductive effects of pressuring children to eat on intake and affect," Appetite, vol. 46, no. 3, pp. 318–323, 2006.

PART V
Policy, Power, and Politics

Food Sovereignty: Power, Gender, and the Right to Food

Rajeev C. Patel

12.1 POWER OVER FOOD

One of the most enduring misconceptions about hunger is that it is primarily the result of a deficit in global food production. If this were so, we might expect food to be absent at times and in places where people die of hunger. Yet economist Amartya Sen has shown that in the majority of cases of widespread famine-related death since WWII, food has been available within the famine-affected area. People have died not for want of food, but for want of the entitlement to eat it [1]. Questions about hunger and its attendant pathologies, therefore, ought to begin with questions about social and political configurations around power over food, rather than about the mere presence or absence of food in the vicinity of a hungry individual.

Although no single commonly agreed definition of hunger exists, two common standards prevail: "undernourishment" and "food security." The former refers to the number of people "whose dietary energy consumption is continuously below a minimum dietary energy requirement for maintaining a healthy life and carrying out a light physical activity" [2]. Undernourishment is a condition suffered by individuals. It is, however, usually established not through individual surveys but through an analysis of a country's food availability, household purchasing power, and entitlements [3],[4]. Current estimates put the world-wide number of undernourished people at nearly one billion [3].

The concept of "food security" attempts to capture the notion of hunger as a deficit not of calories, but as a violation of a broader set of social, economic, and physical conditions. In 1996, the Food and Agriculture Organization of the United Nations (FAO) established at its World Food Summit the most widely agreed definition [5] that "Food security, at the individual, household, national, regional and global levels [is achieved] when all people, at all times, have physical and economic access to sufficient, safe and nutritious food to meet their dietary needs and food preferences for an active and healthy life."

By definition, more people are food insecure than are undernourished, and food insecurity precedes undernourishment. Although there are few people in the United States whose calorie intake is continuously below the threshold of a maintaining healthy life, there are many who, at some point during any given year, are unable to meet their food needs. According to the United States Department of Agriculture (USDA), in 2010 there were 48.8 million US citizens living in food-insecure households. The distribution of food insecurity is uneven. In the US, 21.6 million children lived in food-insecure households, and 35.1% of all female-headed households were food insecure in 2010, compared to 25.4% of male-headed households [6].

Since food insecurity is a broader measure than that of undernourishment, it has been correlated both with hunger and obesity, particularly among women [7]. If hunger is a symptom of a lack of control over the socioeconomic context in which one attempts to eat, it is not unreasonable to understand that lack of control as correlated with factors associated with obesity too. It is possible to have sufficient calories, but insufficiently nutritious food for a healthy life. Armed with this understanding, and with persistent evidence across countries of women and girls' disempowerment compared to men and boys [8], it becomes easier to appreciate the systematically higher rates of food insecurity among women.

12.2 GENDER AND FOOD

The link between gender and food becomes clearer through an understanding of power and control in the food system. Giving away food does little to address the underlying causes of disempowerment that lead to hunger [9]. One group that has articulated this is an international peasant movement called La Via Campesina (see Box 1). They argue that if governments aim merely for food security as a policy goal, the politically difficult questions of inequality in power that produced food insecurity would be ignored, and a broken system would be patched up with entitlements [1]. It is possible, after all, to be food secure in prison where one might continually access safe and nutritious food, yet remain fundamentally disempowered over the process and politics of the food's production, consumption, and distribution.

BOX 1. LA VIA CAMPESINA

La Via Campesina is an organization of farmers, peasants, and landless workers' movements with over 150 million combined members in 70 countries [46]. Its first meeting was held in 1993, and it was constituted as an umbrella organization for a range of social movements that had, through the 1980s, begun to work more closely in Asia, the Americas, and Europe. These movements had come into contact with one another through their attempts to understand, resist, and offer alternatives to "free market" agricultural trade. Even before the organization was officially created, La Via Campesina's member organizations had undertaken a range of actions to confront what they saw as inequality in power within the food system. In India, 200,000 farmers protested the patenting of seeds by multinational corporations. In Europe, 30,000 farmers marched on Brussels to offer an alternative policy goal to the achievement of food security. In Brazil, hundreds of thousands of people occupied farmland, upon which they built thriving communities. In 1996, at the same World Food summit at which the most recent definition of food security was written, La Via Campesina codified its vision for an alternative food system under the rubric of "food sovereignty." At a 2009 La Via Campesina meeting, one of the slogans offered by the assembly was that "food sovereignty is an end to all forms of violence against women."

Instead of food security, Via Campesina has advocated for "food sovereignty." Just like the definition of food security, food sovereignty is an evolving and many-faceted term, but it has an invariant core: "communities have the right to define their own food and agriculture policy" [10]. To be clear, sovereignty is not a call for self-sufficiency, for states to grow within their borders sufficient food to feed their citizens. La Via Campesina instead calls for people to be sovereign over their food systems, for people to have the power to decide what the system should look like. This is an intentionally vague call, with many questions left open-ended, so that the communities involved in claiming food sovereignty might answer issues around production, distribution, and consumption of food for themselves. It is through food sovereignty, La Via Campesina argues, that food security might be achieved, and undernourishment eradicated.

The main demand in food sovereignty is that, for the first time, decisions about the shape of the food system ought to be in the hands not of powerful corporations or geopolitically dominant governments [11], but up to the people who depend on the food system. For the discussion to be representative of the community's desires, however, a non-negotiable element of food sovereignty is women's rights. In order for a democratic conversation about food and agriculture policy to happen, women need to be able to participate in the discussion as freely as men.

Peasant movements, and those who support them, have been castigated as romantics pining for an unattainable past [12]. An insistence of women's rights places food sovereignty firmly in the twenty-first century. This has a practical purpose. Of those undernourished, 60% are women or girls [13]. It is hard to conceive a discussion about hunger without connecting the epidemiology of hunger to women's disempowerment.

On the production side of the food system, women constitute 43% of the agricultural workforce, more often involved in producing food for domestic consumption than export. They are discriminated against in issues ranging from land tenure to wages, from government support to access to technology. The FAO notes that "if women had the same access to productive resources as men, they could increase yields on their farms by 20–30 percent. This could raise total agricultural output in developing countries by 2.5–4 percent, which could in turn reduce the number of hungry people in the world by 12–17 percent" [14].

In addition, women stand to bear a disproportionate burden of the consequences of the twenty-first century's predicted global increase in non-communicable disease (NCD) prevalence. In South Asia, for example, NCDs are projected to account for 72% of deaths by 2030, up from 51% in 2008. In Sub-Saharan Africa, the estimates are 46%, up from 28% over the same period [15]. In addition to the duties of paid work, women bear a disproportionate burden of care work in the management of morbidity associated with NCDs [16],[17], especially in contexts of poverty [18]. These are the kinds of inequities to which food sovereignty calls attention.

12.3   SYSTEMIC INEQUITY AND THE RIGHT TO FOOD

Beyond an examination of the inequitable distribution of power at a household level, food sovereignty suggests an investigation of power relations at meso- and macroeconomic levels. La Via Campesina members are, for example, concerned about corporate power within the global economy. The food system's dysfunction continues to be lucrative for a range of food and agriculture companies. Profits often derive from the increased consumption of processed food, which in turn have driven a global obesity epidemic. Yet the distribution mechanisms within the food system that ration food on the basis of ability to pay have produced the paradox of a billion hungry during a time when there are more than 1.5 billion people overweight [19],[20].

Within the food system, power is concentrated in the hands of a few corporations. In 2008, the top ten agrochemical corporations controlled almost 90% of the global sales of pesticides. Of the US$22 billion global proprietary seed market, only ten corporations controlled 67% [21]. In 2005, the top four beef packing firms controlled 83.5% of the market in the US [22], and worldwide, 40% of all groceries were sold by only 100 retailers [21]. These trends across the food industry have been on an almost-steady climb since they were recorded first in the 1970s. As the US government recently found, "for example, in the pork sector, the market share of the largest four hog slaughtering firms increased from 36 percent in 1982 to 63 percent in 2006. In addition, at the retail level, the share of grocery store sales held by the largest four firms more than doubled, from 16 percent in 1982 to 36 percent in 2005" [23].

This concentration of power has gendered consequences. In contexts where women perform the majority of horticultural and agronomic innovation, they can find their agroecological knowledge supplanted by the technologies of industrial agriculture. Pesticide companies own the largest seed companies, and their agricultural model, dependent on purchased supplies of hybrid seeds and chemical inputs, favors larger, more capital-intensive farms. Women have systematically less access to both land and capital than men, and despite an often sophisticated level of knowledge about farming systems, women's views seldom matter in the shaping of choices around agricultural technologies and food policy [24]. In addition, employment within agriculture consistently pays women around 25% less than men. When food is accessed through market mechanisms, this increases women's systemic risk of hunger [25].

It is for these reasons that women leaders within peasant movements have taken strong stands against multinational corporations such as Monsanto and

Cargill [26]. To be sure, concentration of agricultural power is not new. At the turn of the nineteenth century, four firms—Dreyfus, Cargill, Continental, and Bunge—dominated global grain trading [27]. Today, however, the extent to which food markets matter is far greater. Agricultural market concentration is evident not only in international trade, but across domestic production, distribution, and consumption. This concentration matters more when there are fewer alternatives to the markets in which concentration occurs.

12.4   THE ROLE OF MARKETS AND GOVERNMENTS

To understand why the private sector has achieved such power, it is worth looking at other actors' roles within the food system. Philanthropic foundations have, for example, been responsible for advancing the kinds of industrial agriculture that has imperiled La Via Campesina's members [28],[29]. The "Green Revolution," in which farmers were encouraged and sometimes forced by governments to adopt a system of farming involving hybrid seeds, fertilizer, and pesticides, was initially funded by the Rockefeller and Ford Foundations, and is currently being encouraged by the Gates Foundation in Africa [30],[31],[32]. These farming systems have had gender-negative impacts, as women's knowledge is excluded, and women are systematically less able to control the capital required to participate in resource-intensive farming [33],[34],[35].

National governments and international organizations have also been faulted for their behavior in shaping the food system. Of particular interest to La Via Campesina is the extent to which, through international economic agreements such as the World Trade Organization's (WTO's) Agreement on Agriculture, governments have enabled private sector markets to expand their influence within the food system. A central demand in La Via Campesina's call for food sovereignty is for the WTO to "get out of agriculture" [36]. By this they mean not only ought the Agreement on Agriculture within the WTO be nullified, but that a range of other WTO provisions that affect agriculture, such as rules on intellectual property rights on seeds and phytosanitary measures, also be suspended. Trade agreements rules are influenced by the corporations that subsequently benefit from them [37], with demonstrated gendered impacts as a result [38],[39].

Food corporations continue to attempt to shape domestic and international public policy. PepsiCo, for instance, has gone to great lengths to claim a place at the table in addressing public health issues [40]. Yet the company has since 2000 spent US$26.88 million on lobbying in the US [41], in particular in response to taxes on its products and voicing its concerns on restrictions on marketing its foods to children [42],[43]. PepsiCo's behavior is emblematic of a wider trend in private sector spending within the food system. In a context of shrinking public budgets, and the transformation of public institutions such as schools into sites for the sale of obesogenic products [44], the influence of private interest in public policy matters immensely. Yet the food industry is pushing public debate toward an interpretation of the rise of NCDs as fundamentally a problem of individuals [45]. To accept this is to urge a policy response in which NCDs can be remedied by better individual behavior, rather than more regulation. With women more responsible than men for children's diets, this has the effect of pathologizing individual women, rather than finding fault with a system that removes their freedom to make their children's diets healthier.

12.5 CONCLUSION

The inequalities in power that characterize the food system can be found in households, corporations, regional and state governments, private philanthropic foundations, and international organizations. The strengths of a food sovereignty approach lie in the heuristic approach to power relations that it invites, particularly with respect to gender. For La Via Campesina, and many others, identifying inequities in power within the global food system is more than an academic exercise—it is a means not only to interpret the system, but also to change it.

REFERENCES

1. Sen AK (1981) Poverty and famines: An essay on entitlement and deprivation. New York: Oxford University Press.
2. Food and Agriculture Organization of the United Nations (2011) FAOSTAT Glossary. Rome: Food and Agricultural Organization of the United Nations. Accessed 18 January 2012.
3. Food and Agriculture Organization of the United Nations (2011) The state of food insecurity in the world: How does international price volatility affect domestic economies and food security? Rome: Food and Agricultural Organization of the United Nations. Accessed: 18 January 2012.
4. Food and Agriculture Organization of the United Nations (2004) The state of food insecurity in the world 2004: Monitoring progress towards the World Food Summit and Millennium Development Goals. Rome: Food and Agricultural Organization of the United Nations.

5. Food and Agriculture Organization of the United Nations (2003) Trade reforms and food security: Conceptualising the linkages. Rome: Commodity Policy and Projections Service, Commodities and Trade Division. Accessed 18 January 2012.

6. Coleman-Jensen A, Nord M, Andrews M, Carlson S (2011) Household food security in the United States in 2010. Washington (D.C.): United States Department of Agriculture. Economic Research Report Number 125. Accessed 18 January 2012.

7. Larson NI, Story MT (2011) Food insecurity and weight status among U.S. children and families: A review of the literature. Am J Prev Med. 40. : 166–173. doi:10.1016/j.amepre.2010.10.028.

8. United Nations (2009) World survey on the role of women in development. In: Secretary-General Rot , editor. Accessed 18 January 2012.

9. Rosset P (2011) Preventing hunger: Change economic policy. Nature. 479. : 472–473. doi:10.1038/479472a.

10. Patel R (2009) What does food sovereignty look like? J Peasant Stud 36: 663–673.

11. View Article PubMed/NCBI Google Scholar

12. Patel R (2007) Stuffed and starved: Markets, power and the hidden battle for the world food system. London: Portobello Books.

13. Collier P (2008) The politics of hunger: How illusion and greed fan the food crisis. Foreign Affairs 87. Accessed 18 January 2012.

14. WFP World Food Programme (2009) WFP Gender policy and strategy: Promoting gender equality and the empowerment of women in addressing food and nutrition challenges. Rome: World Food Programme. Accessed 18 January 2012.

15. Food and Agriculture Organization of the United Nations (2011) The State of Food and Agriculture 2010–2011: Women in agriculture - Closing the gender gap for development. Rome: Food and Agricultural Organization of the United Nations.

16. World Bank (2011) The Growing danger of non-communicable diseases: Acting now to reverse course. Washington (D.C.): World Bank Human Development Network. Accessed 18 January 2012.

17. Kramer BJ, Kipnis S (1995) Eldercare and work-role conflict: Toward an understanding of gender differences in caregiver burden. Gerontologist 35: 340–348.

18. Pinquart M, Sörensen S (2006) Gender differences in caregiver stressors, social resources, and health: An updated meta-analysis. J Gerontol B Psychol Sci Soc Sci 61: P33–P45.

19. Kipp W, Tindyebwa D, Rubaale T, Karamagi E, Bajenja E (2007) Family caregivers in rural Uganda: The hidden reality. Health Care Women Int. 28. : 856–871. doi:10.1080/07399330701615275.

20. Swinburn BA, Sacks G, Hall KD, McPherson K, Finegood DT, et al. (2011) The global obesity pandemic: Shaped by global drivers and local environments. Lancet. 378. : 804–814. doi:10.1016/S0140-6736(11)60813-1.

21. World Health Organization (2011) Obesity and overweight. World Health Organization. Accessed 18 January 2012.

22. ETC Group (2008) Who owns nature? Corporate power and the final frontier in the commodification of life. Winnipeg, MB: ETC Group. Accessed 18 January 2012.

23. Hendrickson M, Heffernan WD (2007) Concentration of agricultural markets. Washington (D.C.): National Farmers Union. Accessed 18 January 2012.

24. Government Accountability Office (2009) U.S. Agriculture: Retail food prices grew faster than the prices farmers received for agricultural commodities, but economic research has

not established that concentration has affected these trends [memo]. Washington (D.C.): Government Accountability Office. Accessed 18 January 2012.

25. Feldman S, Welsh R (1995) Feminist knowledge claims, local knowledge, and gender divisions of agricultural labor: Constructing a successor science. Rural Sociol 60: 23–43. doi:10.1111/j.1549-0831.1995.tb00561.x.

26. Hertz T, Winters P, de la O AP, Quiñones EJ, Davis B, et al. (2008) Wage inequality in international perspective: Effects of location, sector, and gender. Rome: Food and Agriculture Organization of the United Nations. Accessed 18 January 2012.

27. Chacko S (2001) Changing the stream: Backgrounder on the women's movement in India. Bangalore: Centre for Education and Documentation.

28. Murphy S (2006) Concentrated market power and agricultural trade: Ecofair trade discussion paper 1. Berlin: Heinrich Boell Stiftung. Accessed 18 January 2012.

29. Jennings BH (1988) Foundations of international agricultural research: Science and politics in Mexican agriculture. Boulder: Westview Press.

30. Cullather N (2010) The hungry world: America's Cold War battle against poverty in Asia. Cambridge (MA): Harvard University Press.

31. Shiva V (1989) The violence of the Green Revolution. Ecological degradation and political conflict in Punjab. Dehra Dun: Research Foundation for Science and Ecology.

32. Perkins JH (1997) Geopolitics and the green revolution: Wheat, genes and the cold war. Oxford: Oxford University Press.

33. Dowie M (2001) American foundations: An investigative history. Cambridge (MA): MIT Press.

34. Sobha I (2007) Green revolution: Impact on gender. J Hum Ecol 22: 107–113. Accessed 18 January 2012.

35. Hart G (1992) Household production reconsidered: Gender, labor conflict, and technological change in Malaysia's Muda region. World Dev 20: 809–823.

36. Cain ML (1981) Java, Indonesia: The introduction of rice processing technology. In: Dauber R, Cain ML, editors. Boulder, Colorado: Westview Press.

37. La Via Campesina (1999) Seattle declaration: Take WTO out of agriculture. Seattle WA: La Via Campesina. Accessed 18 January 2012.

38. Love R (2007) Corporate wealth or public health? WTO/TRIPS flexibilities and access to HIV/AIDS antiretroviral drugs by developing countries. Development in practice. 17. : 208–219. doi:10.1080/09614520701195915.

39. Paul P, Mukhopadhyay K (2010) Growth via intellectual property rights versus gendered inequity in emerging economies: An ethical dilemma for international business. J Bus Ethics. 91. : 359–378. doi:10.1007/s10551-009-0088-y.

40. Çagatay N (2001) Trade, gender and poverty. New York: United Nations Development Programme. Accessed 18 January 2012.

41. Yach D (2011) The critical role of the food industry in the obesity debate. Purchase NY: Pepsico Inc. Accessed 18 January 2012.

42. Center for Responsive Politics (2011) Lobbying spending database: PepsiCo Inc 2011. Washington (D.C.): Center for Responsive Politics/Senate Office of Public Records. Accessed 18 January 2012.

43. Young J (2009) Coke, Pepsi step up spending after being targeted by healthcare reform tax. The Hill. Accessed 18 January 2012.

44. Pepsico Inc (2011) Lobbying Disclosure Act of 1995 (Section 5) Lobbying Report. Senate of the United States. Washington (D.C.): Office of Public Records. Accessed 18 January 2012.

45. Carter M-A, Swinburn B (2004) Measuring the 'obesogenic' food environment in New Zealand primary schools. Health Promot Int. 19. : 15–20. doi:10.1093/heapro/dah103.

46. Jenkin GL, Signal L, Thomson G (2011) Framing obesity: The framing contest between industry and public health at the New Zealand inquiry into obesity. Obesity Rev 12: 1055–1030. doi:10.1111/j.1467-789X.2011.00918.x.

47. Martínez-Torres ME, Rosset PM (2010) La Vía Campesina: The birth and evolution of a transnational social movement. J Peasant Stud. 37. : 149–175. doi:10.1080/03066150903498804.

Big Food, Food Systems, and Global Health

David Stuckler and Marion Nestle

13.1 INTRODUCTION

As the *PLoS Medicine* series on Big Food (www.ploscollections.org/bigfood) kicks off, let's begin this essay with a blunt conclusion: Global food systems are not meeting the world's dietary needs [1]. About one billion people are hungry, while two billion people are overweight [2]. India, for example, is experiencing rises in both: since 1995 an additional 65 million people are malnourished, and one in five adults is now overweight [3],[4]. This coexistence of food insecurity and obesity may seem like a paradox [5], but over- and undernutrition reflect two facets of malnutrition [6]. Underlying both is a common factor: food systems are not driven to deliver optimal human diets but to maximize profits. For people living in poverty, this means either exclusion from development (and consequent food insecurity) or eating low-cost, highly processed foods lacking in nutrition and rich in sugar, salt, and saturated fats (and consequent overweight and obesity).

To understand who is responsible for these nutritional failures, it is first necessary to ask: *Who rules global food systems?* By and large it's "Big Food," by which we refer to multinational food and beverage companies with huge and concentrated market power [7],[8]. In the United States, the ten largest food companies control over half of all food sales [9] and worldwide this proportion is about 15% and rising. More than half of global soft drinks are produced by large multinational companies, mainly Coca-Cola and PepsiCo [10]. Three-fourths of world food sales involve processed foods, for which the largest manufacturers hold over a third of the global market [11]. The world's food system is not a competitive marketplace of small producers but an oligopoly. What people eat is increasingly driven by a few multinational food companies [12].

Virtually all growth in Big Food's sales occurs in developing countries [13] (see Figure 1). The saturation of markets in developed countries [14], along with the lure of the 20% of income people spend on average on food globally, has stimulated Big Food to seek global expansion. Its rapid entry into markets in low- and middle-income countries (LMICs) is a result of mass-marketing campaigns and foreign investment, principally through takeovers of domestic food companies [15]. Trade plays a minimal role and accounts for only about 6% of global processed food sales [15]. Global producers are the main reason why the "nutrition transition" from traditional, simple diets to highly processed foods is accelerating [16],[17].

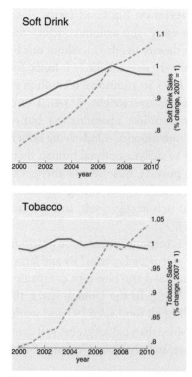

FIGURE 13.1   Growth of Big Food and Big Tobacco sales in developing countries: An example. Shaded blue line is developed countries, dashed grey line is developing countries. Source: Passport Global Market Information Database: EuroMonitor International, 2011 [12].

Big Food is a driving force behind the global rise in consumption of sugar-sweetened beverages (SSBs) and processed foods enriched in salt, sugar, and fat [13]. Increasing consumption of Big Food's products tracks closely with rising levels of obesity and diabetes [18]. Evidence shows that SSBs are major contributors to childhood obesity [19],[20], as well as to long-term weight-gain, type 2 diabetes, and cardiovascular disease [21],[22]. Studies also link frequent consumption of highly processed foods with weight gain and associated diseases [23].

Of course, Big Food may also bring benefits—improved economic performance through increased technology and know-how and reduced risks of undernutrition—to local partners [24]. The extent of these benefits is debatable, however, in view of negative effects on farmers and on domestic producers and food prices [25].

13.2   PUBLIC HEALTH RESPONSE TO BIG FOOD: A FAILURE TO ACT

Public health professionals have been slow to respond to such nutritional threats in developed countries and even slower still in developing countries. Thanks to insights from tobacco company documents, we have learned a great deal about how this industry sought to avoid or flout public health interventions that might threaten their profits. We now have considerable evidence that food and beverage companies use similar tactics to undermine public health responses such as taxation and regulation [26],[27],[28],[29], an unsurprising observation given the flows of people, funds, and activities between Big Tobacco and Big Food. Yet the public health response to Big Food has been minimal.

We can think of multiple reasons for the failure to act [30]. One is the belated recognition of the importance of obesity to the burden of disease in LMICs [13]. The 2011 Political Declaration of the United Nations High-Level Meeting on Prevention and Control of Non-communicable Diseases (NCDs) recognized the urgent case for addressing the major avoidable causes of death and disability [31], but did not even mention the roles of agribusiness and processed foods in obesity. Despite evidence to the contrary, some development agencies continue to view obesity as a "disease of affluence" and a sign of progress in combating undernutrition [32].

A more uncomfortable reason is that action requires tackling vested interests, especially the powerful Big Food companies with strong ties to and influence over national governments. This is difficult terrain for many public health scientists. It took five decades after the initial studies linking tobacco and cancer for effective public health policies to be put in place, with enormous cost to human health. Must we wait five decades to respond to the similar effects of Big Food?

If we are going to get serious about such nutritional issues, we must make choices about how to engage with Big Food. Whether, and under what circumstances, we should view food companies as "partners" or as part of the solution to rising rates of obesity and associated chronic diseases is a matter of much current debate, as indicated by the diverse views of officials of PepsiCo and nutrition scientists [24],[27],[28],[33],[34].

13.3   ENGAGING WITH BIG FOOD—THREE VIEWS

We see three possible ways to view this debate. The first favors voluntary self-regulation, and requires no further engagement by the public health community. Those who share this view argue that market forces will self-correct the negative externalities resulting from higher intake of risky commodities. Informed individuals, they say, will choose whether to eat unhealthy foods and need not be subjected to public health paternalism. On this basis, UN secretary-general Ban Ki Moon urged industry to be more responsible: "I especially call on corporations that profit from selling processed foods to children to act with the utmost integrity. I refer not only to food manufacturers, but also the media, marketing and advertising companies that play central roles in these enterprises" [35]. Similarly, the UK Health Minister recently said: "the food and drinks industry should be seen, not just as part of the problem, but part of the solution… An emphasis on prevention, physical activity and personal and corporate responsibility could, alongside unified Government action, make a big difference" [36].

The second view favors partnerships with industry. Public health advocates who hold this view may take jobs with industry in order to make positive changes from within, or actively seek partnerships and alliances with food companies. Food, they say, is not tobacco. Whereas tobacco is demonstrably harmful in all forms and levels of consumption, food is not. We can live without tobacco, but we all must eat. Therefore, this view holds that we must work with Big Food to make healthier products and market them more responsibly.

The third approach is critical of both. It recognizes the inherent conflicts of interest between corporations that profit from unhealthy food and public health collaborations. Because growth in profit is the primary goal of corporations, self-regulation and working from within are doomed to fail. Most proponents of this viewpoint support public regulation as the only meaningful approach, although some propose having public health expert committees set standards and monitor industry performance in improving the nutritional quality of food products and in marketing the products to children.

We support the critical view, for several reasons. First, we find no evidence for an alignment of public health interest in curbing obesity with that of the food and beverage industry. Any partnership must create profit for the industry, which has a legal mandate to maximize wealth for shareholders. We also see no obvious, established, or legitimate mechanism through which public health professionals might increase Big Food's profits.

Big Food attains profit by expanding markets to reach more people, increasing people's sense of hunger so that they buy more food, and increasing profit margins through encouraging consumption of products with higher price/cost surpluses [28]–[31],[37]. Industry achieves these goals through food processing and marketing, and we are aware of no evidence for health gains through partnerships in either domain. Although in theory minimal processing of foods can improve nutritional content, in practice most processing is done so to increase palatability, shelf-life, and transportability, processes that reduce nutritional quality. Processed foods are not necessary for survival, and few individuals are sufficiently well-informed or even capable of overcoming marketing and cost hurdles [38]. Big Food companies have the resources to recruit leading nutritional scientists and experts to guide product development and reformulation, leaving the role of public health advisors uncertain.

To promote health, industry would need to make and market healthier foods so as to shift consumption away from highly processed, unhealthy foods. Yet, such healthier foods are inherently less profitable. The only ways the industry could preserve profit is either to undermine public health attempts to tax and regulate or to get people to eat more healthy food while continuing to eat profitable unhealthy foods [33],[39]. Neither is desirable from a nutritional standpoint. Whereas industry support for research might be seen as one place to align interests, studies funded by industry are 4- to 8-fold more likely to support conclusions favorable to the industry [40].

Our second reason to support the critical view has to do with the "precautionary principle" [41]. Because it is unclear whether inherent conflicts of interest can be reconciled, we favor proceeding on the basis of evidence. As George Orwell put it, "saints should always be judged guilty until they are proved innocent." We believe the onus of proof is on the food industry. If food companies can rigorously and independently establish self-regulation or private–public partnerships as improving both health and profit, these methods should be extended and replicated. But to date self-regulation has largely failed to meet stated objectives [42],[43],[44],[45],[46],[47], and instead has resulted in significant pressure for public regulation. Kraft's decision to ban trans fats, for example, occurred under pressure of lawsuits [48]. If industry believed that self-regulation would increase profit, it would already be regulating itself.

We believe the critical view has much to offer. It is a model of dynamic and dialectic engagement. It will increase pressures on industry to improve health performance, and it will encourage those who are sympathetic to the first or second views to effect change from within large food and beverage companies.

Public health professionals must recognize that Big Food's influence on global food systems is a problem, and do what is needed to reach a consensus about how to engage critically. The Conflicts of Interest Coalition, which emerged from concerns about Big Food's influence on the U.N. High-Level Meeting on NCDs, is a good place to start [29],[49]. Public health professionals must place as high a priority on nutrition as they do on HIV, infectious diseases, and other disease threats. They should support initiatives such as restrictions on marketing to children, better nutrition standards for school meals, and taxes on SSBs. The central aim of public health must be to bring into alignment Big Food's profit motives with public health goals. Without taking direct and concerted action to expose and regulate the vested interests of Big Food, epidemics of poverty, hunger, and obesity are likely to become more acute.

REFERENCES

1. De Schutter O (2011) Report submitted by the Special Rapporteur on the right to food. Geneva: United Nations. Available: http://www2.ohchr.org/english/issues/food/docs/A-HRC-16-49.pdf.
2. Patel R (2008) Stuffed and starved: The hidden battle for the world food system: Melville House. 448 p.
3. Doak C, Adair LS, Bentley M (2005) The dual burden household and nutrition transition paradox. Int J Obesity 29: 129–136.
4. Stein AD, Thompson AM, Waters A (2005) Childhood growth and chronic disease: evidence from countries undergoing the nutrition transition. Matern Child Nutr 1: 177–184. Available: http://www.ncbi.nlm.nih.gov/entrez/query.fcgi?cmd=Retrieve&db=PubMed&dopt=Citation&list_uids=16881898.
5. Caballero B (2005) A nutrition paradox – underweight and obesity in developing countries. N Engl J Med 352: 1514–1516.
6. Eckholm E, Record F (1976) The two faces of malnutrition. Worldwatch. Available: http://www.worldwatch.org/bookstore/publication/worldwatch-paper-9-two-faces-malnutrition.

7. Pollan M (2003) The (agri)cultural contradictions of obesity. New York Times. Available: http://www.nytimes.com/2003/10/12/magazine/12WWLN.html.

8. Brownell K, Warner KE (2009) The perils of ignoring history: Big Tobacco played dirty and millions died. How similar is Big Food? Milbank Quarterly 87: 259–294.

9. Lyson T, Raymer AL (2000) Stalking the wily multinational: power and control in the US food system. Agric Human Values 17: 199–208.

10. Alexander E, Yach D, Mensah GA (2011) Major multinational food and beverage companies and informal sector contributions to global food consumption: Implications for nutrition policy. Global Health 7: 26.

11. Alfranca O, Rama R, Tunzelmann N (2003) Technological fields and concentration of innovation among food and beverage multinationals. International Food and Agribusiness Management Review 5.

12. EuroMonitor International (2011) Passport Global Market Information Database: EuroMonitor International.

13. Stuckler D, McKee M, Ebrahim S, Basu S (2012) Manufacturing Epidemics: The Role of Global Producers in Increased Consumption of Unhealthy Commodities Including Processed Foods, Alcohol, and Tobacco. PLoS Med. 6. doi:10.1371/journal.pmed.1001235.

14. Hawkes C (2002) Marketing activities of global soft drink and fast food companies in emerging markets: A review. Geneva: World Health Organization. Available: http://www.who.int/hpr/NPH/docs/globalization.diet.and.ncds.pdf.

15. Regmi A, Gehlhar M (2005) Processed food trade pressured by evolving global supply chains. Amberwaves: US Department of Agriculture. Available: http://www.ers.usda.gov/amberwaves/february05/features/processedfood.htm.

16. Popkin B (2002) Part II: What is unique about the experience in lower- and middle-income less-industrialised countries compared with the very-high income countries? The shift in the stages of the nutrition transition differ from past experiences! Public Health Nutr. 5. : 205–214. doi:10.1079/PHN2001295.

17. Hawkes C (2005) The role of foreign direct investment in the nutrition transition. Public Health Nutri 8: 357–365.

18. Basu S, Stuckler, D McKee M, Galea G (2012) Nutritional drivers of worldwide diabetes: An econometric study of food markets and diabetes prevalence in 173 countries. Public Health Nutrition. In press.

19. Maliv V, Schulze MB, Hu FB (2006) Intake of sugar-sweetened beverages and weight gain: A systematic review. Am J Clin Nutr 84: 274–288.

20. Moreno L, Rodriguez G (2007) Dietary risk factors for development of childhood obesity. Curr Opin Clin Nutr Metab Care 10: 336–341.

21. Hu F, Malik VS (2010) Sugar-sweetened beverages and risk of obesity and type 2 diabetes. Physiol Behav 100: 47–54.

22. Malik V, Popkin BM, Bray GA, Despres JP, Hu F (2010) Sugar-sweetened beverages, obesity, type 2 diabetes mellitus, and cardiovascular disease risk. Circulation 121: 1356–1364.

23. Pereira M, Kartashov AI, Ebbeling CB, Van Horn L, Slattery ML, et al. (2005) Fast food habits, weight gain and insulin resistance in a 15-year prospective analysis of the CARDIA study. Lancet 365: 36–42.

24. Yach D, Feldman ZA, Bradley DG, Khan M (2010) Can the food industry help tackle the growing burden of undernutrition? Am J Public Health 100: 974–980.

25. Evenett S, Jenny F (2011) Trade, competition, and the pricing of commodities. Washington D.C.: Center for Economic Policy Research. Available: http://www.voxeu.org/reports/CEPR-CUTS_report.pdf.

26. Chopra M, Darnton-Hill I (2004) Tobacco and obesity epidemics: Not so different after all? BMJ 328: 1558–1560.

27. Ludwig D, Nestle M (2008) Can the food industry play a constructive role in the obesity epidemic? JAMA 300: 1808–1811.

28. Wiist W (2011) The corporate playbook, health, and democracy: The snack food and beverage industry's tactics in context. In: Stuckler D, Siegel , K , editors. Oxford: Oxford University Press.

29. Stuckler D, Basu S, McKee M (2011) UN high level meeting on non-communicable diseases: An opportunity for whom? BMJ. 343. d5336 p. doi:10.1136/bmj.d5336.

30. Stuckler D (2008) Population causes and consequences of leading chronic diseases: A comparative analysis of prevailing explanations. Milbank Quarterly 86: 273–326.

31. UN General Assembly (2011) Political declaration of the High-level Meeting of the General Assembly on the Prevention and Control of Non-communicable Diseases (NCDs). New York: UN. Available: http://www.un.org/en/ga/ncdmeeting2011/.

32. Mitchell A (2011) Letter to National Heart Forum about 'Priority actions for the NCD crisis'. In: Lincoln P, editor. London: UK DFID.

33. Monteiro C, Gomes FS, Cannon G (2009) The snack attack. Am J Public Health 100: 975–981.

34. Acharya T, Fuller AC, Mensah GA, Yahc D (2011) The current and future role of the food industry in the prevention and control of chronic diseases: The case of PepsiCo. In: Stuckler D, Siegel , K , editors. Oxford: Oxford University Press.

35. Ki-Moon B (2011) Remarks to the General Assembly meeting on the prevention and control of non-communicable disease. Geneva: UN. Available: http://www.un.org/apps/news/infocus/sgspeeches/statments_full.asp?statID=1299.

36. Lansley A (2011) 4th plenary meeting. Geneva: UN. Available: http://www.ncdalliance.org/sites/default/files/rfiles/Monday%20Sep%2019%203pm.pdf.

37. Koplan J, Brownell KD (2010) Response of the food and beverage industry to the obesity threat. JAMA 304: 1487–1488.

38. Wansink B (2007) Mindless eating: Why we eat more than we think. Bantam Books.

39. Wilde P (2009) Self-regulation and the response to concerns about food and beverage marketing to children in the United States. Nutr Rev 67: 155–166.

40. View Article PubMed/NCBI Google Scholar

41. Lesser L, Ebbeling CB, Goozner M, Wypij D, Ludwig DS (2008) Relationship between funding source and conclusion among nutrition-related scientific articles. PLoS Med. 4. e5 p. doi:10.1371/journal.pmed.0040005.

42. Raffensperger C, Tickner J (1999) Protecting public health and the environment: implementing the precautionary principle. Washington D.C.: Island Press.

43. Lewin A, Lindstrom L, Nestle M (2006) Food industry promises to address childhood obesity: Preliminary evaluation. J Public Health Policy 27: 327–348.

44. Lang T (2006) The food industry, diet, physical activity and health: A review of reported commitments and prctice of 25 of the world's largest food companies. London: Oxford Health Alliance.

45. Sharma L, Teret SP, Brownell KD (2010) The food industry and self-regulation: Standards to promote success and to avoid public health failures. Am J Public Health 100: 240–246.

46. Bonell C, McKee M, Fletcher A, Haines A, Wilkinson P (2011) The nudge smudge: misrepresentation of the "nudge" concept in England's public health White Paper. Lancet 377: 2158–2159.

47. Campbell D (2012) High street outlets ignoring guidelines on providing calorie information. The Guardian. London. Available: http://www.guardian.co.uk/business/2012/mar/15/high-street-guidelines-calorie-information.

48. Hawkes C, Harris JL (2011) An analysis of the content of food industry pledges and marketing to children. Public Health Nutr 14: 1403–1414.

49. View Article PubMed/NCBI Google Scholar

50. Zernike K (2004) Lawyers shift focus from Big Tobacco to Big Food. New York Times. New York. Available: http://www.nytimes.com/2004/04/09/us/lawyers-shift-focus-from-big-tobacco-to-big-food.html.

51. Conflicts of Interest Coalition (2011) Statement of Concern.

Keywords

- Nutrient intakes
- Mood disorders
- Food insecurity
- Mental distress
- Pregnant women
- Ethiopia
- Prevention of mother-to-child transmission of HIV
- PMTCT cascade
- HIV infection
- Zimbabwe
- Maternal health
- HIV
- Nutritional counseling
- Lay health workers
- Microfinance
- Agriculture
- Livelihoods
- Intervention

Author Notes

Chapter 2

Authors' Contributions
KMD conducted the study as part of her doctoral studies and was supervised by BJK. KMD drafted the manuscript and BJK helped with revisions of the manuscript. Both authors read and approved the final manuscript.

Acknowledgments
The authors thank their funding source, The Danone Research Institute. The second author also thanks the Alberta Children's Hospital Research Institute for ongoing support. We also acknowledge the assistance of the Mood Disorders Association of British Columbia for providing support staff, office space and assistance with recruitment.

Funding
Financial support for this project was obtained from The Danone Research Institute, which played no role in carrying out the study, analyzing the results, or influencing publication.

Compliance with Ethical Guidelines
All procedures followed were in accordance with the ethical standards of the responsible committee on human experimentation (institutional and national) and with the Helsinki Declaration of 1975, as revised in 2000.

Competing Interests
The authors declare that they have no competing interests.

Informed Consent
Informed consent was obtained from all participants included in the study.

Chapter 3

Acknowledgement
This study was supported by Jimma University, Ethiopia, University of Minnesota and Khat Research Program (KRP), USA. We would also like to thank data collectors, data clerks and study participants.

Competing Interests

We declare that we have no any competing interests.

Authors' Contributions

The analysis is conceived and performed and drafted by M G.J and MN, MA, AL, RH, FL. Both TB and PK assisted the analysis and interpretation of the data. MT, MT, EK, TG, HS, NW, YT participate during protocol development, designing of the study and monitored the data collection. All authors critically reviewed the manuscript. M G. J and MN, RH, TB &PK also reviewed the manuscript after reviewers' comments. All Authors have read and approved the final version of the manuscript. M G. J is responsible for manuscript submission.

Chapter 4

Author Contributions

Conceived and designed the experiments: ACT KJH. Performed the experiments: ACT. Analyzed the data: ACT. Contributed reagents/materials/analysis tools: ACT. Wrote the first draft of the manuscript: ACT KJH. Contributed to the writing of the manuscript: ACT KJH SDW.ICMJE criteria for authorship read and met: ACT KJH SDW. Agree with manuscript results and conclusions: ACT KJH SDW. Assisted in interpretation of the data: ACT KJH SDW. Revised the manuscript for important intellectual content: ACT KJH SDW.

Funding

ACT receives salary support from the Robert Wood Johnson Foundation Health and Society Scholars Program. SDW receives salary support from U.S. National Institute of Health K23 MH079713, the Hellman Family Foundation, and the Burke Family Foundation. The funders had no role in study design, data collection and analysis, decision to publish, or preparation of the manuscript.

Competing Interests

SDW has previously been affiliated with Physicians for Human Rights, an organization that advocates for the protection of internationally guaranteed rights and/or prosecution of those who violate human rights. All authors have declared that no financial conflicts of interest exist.

Chapter 5

Acknowledgments

We are grateful to all of the women and children who participated in the study. We are also indebted to Dr. Maya Petersen, Dr. Mi-Suk Kang Dufour,

Ms. Constancia Watadzaushe, and Mr. Jeffrey Dirawo for their valuable contributions to the impact evaluation. The evaluation of Zimbabwe's Accelerated National PMTCT Program was supported by the Children's Investment Fund Foundation (CIFF). Dr. McCoy is supported by Award Number K01MH094246 from the National Institute of Mental Health. The content is solely the responsibility of the authors and does not necessarily represent the official views of the National Institute of Mental Health or the National Institutes of Health.

Competing Interests

The authors declare that they have no competing interests.

Authors' Contributions

SM RB, NP, and FC collaboratively designed the impact evaluation which was the source of data for this analysis. SM and RB conducted the data analysis, which was iteratively refined after discussion with all authors. SM drafted the initial manuscript and all authors participated in reviewing the draft for intellectual content and assisting with revisions. All authors approved the final version of the manuscript.

Chapter 6

Acknowledgments

The peer counseling intervention pilot study was funded by the National Institute of Mental Health (NIMH) (R34MH084675, PI: Martínez). Preparation of this manuscript was supported in part by RAND Health. Its contents are solely the responsibility of the authors and do not represent the official views of NIMH or RAND. The authors thank Alexandria Smith, formerly of RAND, who merged the datafiles from the field and created the dataset that was used in these analyses. They are also grateful to their close collaborators in Honduras who made the study possible: the 17 peer nutrition counselors, clinic staff at the 14 pilot sites, ASONAPVSIDAH and the Ministry of Health.

Competing Interests

The authors have no competing interests to declare.

Authors' Contributions

HM developed the peer nutritional counseling intervention and trained the peer counselors. KPD, HM, and KP conceptualized and designed the present study to assess impacts of the intervention. HM, HF, and BR led field implementation.

KPD, BH, and MF analyzed the data. KPD, BH, MF, and HM drafted the paper; KP and HF reviewed the paper and provided critical comment. All authors read and approved the final manuscript.

Chapter 7

Funding

Dr. Aranka Anema is supported by the Canadian Institutes of Health Research (CIHR) Vanier Award for doctoral research. Dr. Sheri D. Weiser is supported by a National Institutes of Health (NIH) grant (R01 MH095683-01) and the Burke Family Foundation. Dr. Julio Montaner is supported by the British Columbia Ministry of Health; through an Avant-Garde Award (No. 1DP1DA026182) from the National Institute of Drug Abuse (NIDA), at the NIH; and through a KT Award from CIHR. He has also received financial support from the International AIDS Society, United Nations AIDS Program, World Health Organization, NIH Research-Office of AIDS Research, National Institute of Allergy & Infectious Diseases, The United States President's Emergency Plan for AIDS Relief (PEPfAR), UNICEF, the University of British Columbia, Simon Fraser University, Providence Health Care and Vancouver Coastal Health Authority. He has received grants from Abbott, Biolytical, Boehringer-Ingelheim, Bristol-Myers Squibb, Gilead Sciences, Janssen, Merck and ViiV Healthcare. Dr. Robert S. Hogg has held grant funding in the last five years from the NIH, CIHR, Health Canada, Merck, and Social Sciences and Humanities Research Council of Canada (SSHRC). He has also received funding from Agouron Pharmaceuticals Inc, Boehringer Ingelheim Pharmaceuticals Inc, Bristol-Myers Squibb, GlaxoSmithKline, and Merck Frosst Laboratories for participating in continued medical education programmes. The funders had no role in study design, data collection and analysis, decision to publish, or preparation of the manuscript.

Competing Interests

Julio S. G. Montaner has received educational grants from, served as an ad hoc adviser to or spoken at various events sponsored by Abbott Laboratories, Agouron Pharmaceuticals Inc., Boehringer Ingelheim Pharmaceuticals Inc., Borean Pharma AS, Bristol-Myers Squibb, DuPont Pharma, Gilead Sciences, GlaxoSmithKline, Hoffmann-La Roche, Immune Response Corporation, Incyte, Janssen-Ortho Inc., Kucera Pharmaceutical Company, Merck Frosst Laboratories, Pfizer Canada Inc., Sanofi Pasteur, Shire Biochem Inc., Tibotec Pharmaceuticals Ltd. and Trimeris Inc. There are no patents, products in development or marketed products to declare. This does not alter the authors' adherence to all the PLOS ONE policies on sharing data and materials.

Author Contributions
Contributed specialist knowledge and critical review/feedback on manuscript drafts: SDW JSGM RSH. Conceived and designed the experiments: AA RSH. Analyzed the data: AA KC YC. Wrote the paper: AA.

Chapter 8

Acknowledgments
We thank the Kenyan men and women who are participating in the study. We acknowledge the important logistical support of the KEMRI-UCSF Collaborative Group and especially FACES. We gratefully acknowledge the Director of KEMRI, the Director of KEMRI's Centre for Microbiology Research, and the Nyanza Provincial Ministries of Health for their support in conducting this research. We also thank Beatrice Otieno, Nolline Akuku, Clare Aloo, Jackline Atieno, Perez Kitoto, Mark Matinde, Nicholas Otieno, Pamela Omondi, and Amos Onyang for their important contributions to this research.

Funding
The research described was supported by the National Institutes of Mental Health. The study was funded under grant 1R34MH094215. The funders had no role in data collection and analysis.

Competing Interest
The authors declare that they have no competing interests.

Authors' Contributions
Conceived and designed the experiments: CRC, SDW, EAB. Performed the experiments: EW, RLS. Analyzed the data: FW, SS, CRC, SDW, RLS, AMH, RR, KS, SLD. Wrote the paper: CRC, SW, RLS, EW, AH. Contributed to the writing of the manuscript: CRC, SDW, RLS, EW, EAB, LMB. All authors read and approved the final manuscript.

Chapter 9

Disclosure
The funders had no role in the design or conduct of the study; in the collection, analysis, and interpretation of the data; or in the preparation or approval of the paper.

Disclaimer

All authors, external and internal, had full access to all of the data (including statistical reports and tables) in the study and can take responsibility for the integrity of the data and the accuracy of the data analysis.

Conflict of Interests

There is no potential conflict of interests.

Authors' Contribution

Enza Gucciardi and Margaret DeMelo conceived the study. Enza Gucciardi, Justine Chan, and Margaret DeMelo were involved in designing the study and developing the methods. Justine Chan and Enza Gucciardi obtained funding. Justine Chan coordinated the study and conducted the individual interviews. Justine Chan, Margaret DeMelo, Enza Gucciardi, and Jacqui Gingras read transcripts, came to a consensus on the analytical framework, and contributed to the analysis. Justine Chan drafted the paper. All authors contributed to the interpretation of the analysis and critically revised the paper. Justine Chan is the guarantor.

Acknowledgments

The authors thank Grace Karam for transcribing the interview manuscripts. This study was supported by Ryerson University Faculty of Community Services seed grant awarded to Enza Gucciardi.

Chapter 10

Acknowledgments

The authors would like to thank the promotora-researchers (Maria Davila, Thelma Aguillon, Hilda Maldonado, Maria Garza, and Esther Valdez); the mothers and children who participated in this project; and the data entry team (Jenny Becker Hutchinson, Kelli Gerard, and Leslie Puckett).

This research was supported in part with funding from the Robert Wood Johnson Foundation Healthy Eating Research Program (#66969), National Institutes of Health (NIH)/National Center on Minority Health and Health Disparities (# 5P20MD002295), Cooperative Agreement #1U48DP001924 from the Centers for Disease Control and Prevention (CDC), Prevention Research Centers Program through Core Research Project and Special Interest Project Nutrition and Obesity Policy Research and Evaluation Network, and by USDA RIDGE Program, subaward (#018000-321470-02) through Southern Rural Development Center, Mississippi State University. The content is solely the responsibility of the authors and does not necessarily represent the official views of the Robert Wood Johnson Foundation, NIH, CDC, and USDA-ERS.

Competing Interests
The authors declare that they have no competing interests.

Authors' Contributions
JRS designed the study, and worked on the development of the instrument and the protocol for collection of data. JRS, CN, CMJ, and WRD wrote the first draft of the paper. JRS, CN, CMJ, and WRD read and approved the final manuscript.

Chapter 11

Acknowledgments
This paper was supported by funds from the Robert Wood Johnson Foundation through its Healthy Eating Research program. This project has been funded in part with federal funds from the National Cancer Institute, National Institutes of Health, under contract no. HHSN261200800001E. The content of this publication does not necessarily reflect the views or policies of the Department of Health and Human Services nor does mention of trade names, commercial products, or organizations imply endorsement by the USA Government.

Chapter 12

Acknowledgments
The author gratefully acknowledges invaluable comments from Maninder Kahlon and the research support of Meredith Palmer.

Author Contributions
Wrote the first draft of the manuscript: RP. Contributed to the writing of the manuscript: RP.ICMJE criteria for authorship read and met: RP. Agree with manuscript results and conclusions: RP.

Funding
No specific funding was received to write this article.

Competing Interests
RP is both a Fellow at the Institute for Food and Development Policy, and a Fellow of the Institute for Agriculture and Trade Policy's (IATP) Food And Community Fellowship program. This program is funded, in part, by the W. K. Kellogg Foundation though fellows are appointed by IATP. RP has no relationship with La Via Campesina.

Chapter 13

Author Contributions

Analyzed the data: DS. Wrote the first draft of the manuscript: DS. Contributed to the writing of the manuscript: DS MN. ICMJE criteria for authorship read and met: DS MN. Agree with manuscript results and conclusions: DS MN.

Funding

No specific funding was received for writing this article.

Competing Interests

MN and DS are the guest editors of the PLoS Medicine series on Big Food.

Index

A

Aboriginal ancestry, 109, 111, 113
Academy of Nutrition and Dietetics, 205
Acute/chronic food insecurity, 194
Acute respiratory infection, 93
Adequate macronutrient distribution, 14
Adjusted
  hazard ratio, 113
  odds ratio, 32, 37, 54
Adok Timo, 128, 129, 148
Agriculture, 3, 90, 108, 137, 210, 222–226
  challenges, 147
  technologies, 225
  training, 129, 140, 144
Agrovet, 129
Alcohol Use Disorders Identification Test,
  136
Alimentos
  constructores, 93
  energéticos, 93
  reguladores, 93
American Nurses Association, 205
Analysis of, outcome data, 140
Anemia, 94
Antenatal care (ANC), 29, 30, 37, 39, 69, 70,
  73, 76–81
Anthropometric, 90, 183
Anthropometry, 92, 94, 139
Antiretroviral, 105, 109, 111–115, 126, 127,
  143
  prophylaxis (ARV), 69, 70, 73, 77–81,
    127, 134
  therapy (ART), 69, 70, 73, 77, 78–81,
    89–94, 99–101, 105–107, 109–118,
    126, 130–139, 141, 143, 144, 148, 150
Anxiety, 3, 22, 28, 37, 60, 70, 72, 92, 94, 108,
  116, 117, 132
Appetite, 22, 93

Asociación Nacional de Personas Viviendo
  con VIH/SIDA en Honduras, 91
Assessment of HIV status, 72
Association between food insecurity
  mental distress, 28, 35, 40
  sexual risk outcomes, 55
Attention, 90, 200, 203, 204, 212, 224

B

Barriers to Shamba Maisha implementation,
  146
Baseline characteristics, 110, 111, 141, 187
Big Food, 231–236
Binary logistic regression analyses, 31
Binomial tests, 14
Biologic mechanisms, 117
Bivariate comparison, 111
Body mass index, 14, 51, 81, 95, 97, 109, 112,
  114, 130, 135, 142, 183
  forage-and-sex growth charts, 186
  measurements, 134
Breastfeeding, 70–73, 77, 80–82
British Columbia Centre for Excellence in
  HIV/AIDS, 106
British Columbia Nutrition Survey, 13, 14
Bucket irrigation, 145

C

Calcium, 180, 185–187
Canada Revenue Agency, 109
Canned foods, 168
Carbohydrates, 21, 93, 95, 130, 167, 171
Cardiovascular disease, 100, 194, 233
Census block groups, 181
Centers for Disease Control and Prevention,
  183, 186
Centers for Disease Control-Kenya laboratory,
  134
Child and Adult Care Food Program, 202

Child food insecurity, 195
Child's perspective, 181, 194
Children self-reported food security, 184
Children's food security, 183, 192
Chi-square tests, 15, 17, 140
Chronic illness, 175
CNSTAT review and recommendations, 4
    CNSTAT Panel, 4
COBAS TaqMan HIV viral load platform,
    134, 137
Colonias, 179–182, 193–195
Committee on National Statistics, 3
Community Based Education Training
    Program, 29
Community food justice, 204, 213
Community kitchens, 168, 170, 173, 174
Comparisons of nutrient intakes, 18
Condom, 47–54, 57, 60–62
Confidence intervals, 16, 32, 74, 113, 114, 141
Constructores, 95
Consumption, 8, 28, 106, 137, 140, 142, 162,
    185, 186, 203, 206, 209, 210, 221–226,
    233–235
Continuous variables, 17, 54, 117, 140
Cost-benefit analysis, 129
Covariates, 14, 57, 73, 74, 79, 95, 137
Cronbach's alpha (reliability coefficient), 35,
    53, 183
Crude and adjusted prevalence ratios, 21
Crude odds ratios, 37
Crude prevalence ratios, 17
Cultural factors, 28

D

Data analysis, 14, 81, 167
    procedure, 164
        cool-down/wrap-up, 166
        health services, 165
        income and social status, 165
        personal health practices and coping
            skills, 166
        social environments/social support
            networks, 165
Data collection, 71, 134, 139, 182
Data management and analysis, 31

Debt recoveries, 146
Decision makers, 11
Demographics, 137, 183
Dependent variable, 13, 31, 61
Depression, 11, 13, 17, 22, 28, 37, 38, 60, 61,
    70, 117, 131, 132, 135, 143, 170, 173
Descriptive analyses, 95, 140
Diabetes
    management, 161, 162, 167, 168, 171,
        173, 175
    mellitus, 161
    retinopathy, 173
    self-management, 161, 175
Diarrhea, 92–94, 130
Dietary
    fiber, 180, 185, 186
    folate equivalents, 18, 19
    folate intake, 22
    intake, 12, 92, 94, 95, 97, 99, 118, 142,
        143, 148, 180, 181, 184, 186, 193–195,
        209
    reference intakes, 14
Distribution gap, 8
Drug addiction, 106
Drug treatment program, 106

E

Earned Income Tax Credit, 210
Economic
    factors, 210
    food environment, 210
    indicators, 142
    benefit transfer, 203
Energéticos, 95
Engaging with Big Food—three views, 234
Enhancing agency and resilience, 167, 171
Environmental types, 208
Eroding economic productivity, 126
Escala Brasiliera de Segurança Alimentar, 50
Ethiopia, 27–30, 32, 34, 36–40
Expenditures, 137, 140

F

Factors associated with mental distress, 36

Family AIDS Care & Education Services, 132
Farm life, 127
Farming techniques, 129
Federal and community food and nutrition assistance programs, 195
Federal food assistance programs, 202
Fertility preferences, 52, 57
Fertilization and crop rotation, 129
Fisher's exact test, 110
Flow cytometry, 109
Fluorescent monoclonal, 109
Folate, 12, 13, 17, 18, 21, 22
Food aid and nutrition counseling, 91
Food and Agriculture Organization, 130, 222
Food and Nutrition Technical Assistance, 126
Food assistance initiatives, 201, 206
Food assistance programs, 83, 200–202, 209
Food availability, 221
Food groups, 93, 94, 135
Food insecure parents, 212
Food insecurity, 3–8, 11–23, 28, 31–40, 47–62, 70–83, 89–99, 106–118, 125–133, 136, 143, 150, 162, 163, 167, 168, 170–175, 179–181, 193–195, 199–208, 210–213, 222, 223, 231
    nutrient intakes, 17
    psychological functioning, 15, 17, 22
    psychological issues, 11
        depression, 11
        eating disorders, 11
        impaired cognition, 11
    score, 94
Food insufficiency, 11, 22, 47, 60, 108
Food labeling, 210
Food marketing and advertising, 211
Food policy, 225
Food quality, 72, 94, 213
Food security, 3–8, 11, 18, 23, 40, 57, 60, 70–82, 90, 94, 99–111, 115–118, 127, 130, 131, 139, 140, 142, 148, 162, 174, 181–195, 200, 203–208, 210–213, 221–224
    status, 181, 186, 187, 193, 194
Food sovereignty, 223, 224, 225, 226, 227
Food subsidies, 210
Food supplement benefits, 202
Food systems approach, 204, 206, 213

reducing food insecurity, 204
Foreign-born children/parents, 193

G

Gelberg and Anderson's Behavioral Model, 135
Gender and food, 222
Gender empowerment, 136, 143
Gender inequality, 127
General population sample, 12, 15, 17
Global Assessment of Functioning Scale, 13, 20
Global food
    production, 221
    security, 7, 8
Green revolution, 226
Grocery gateway, 174
Grocery shopping, 167

H

Hail storms, 147
Ham-D scores, 22
Hamilton depression scale, 13, 20
Head start program, 210
Health care services, 135
Health researchers, 11, 202
Health services, 70, 72, 79, 80, 165
Healthy foods, 167, 174, 180, 184
Healthy People 2020 objectives, 203
Heart to Home Meals, 174
Heterogeneity, 149
Hip pump, 127, 129, 140
Hispanic households, 179, 180, 193
Hispanics, 180, 193
HIV, 47–51, 57, 60–62, 69–80, 82, 89–94, 100, 101, 105–118, 125–139, 144, 149, 236
    infected adults, 131
    infection, 77, 78, 79
    positive IDU, 116, 118
    positive individuals, 105, 106, 109, 110, 116–118, 133
    related activities, 149
    related wasting, 106
    side effects, 92
    testing, 61, 69, 73

treatment programs, 118
Honduras, 89, 90, 91, 99
Household
  characteristics with very low food secu-
    rity, 5
  economic indicators, 140, 148, 150
  food insecurity, 162
    access prevalence, 31
    access scale, 31, 35, 72, 76, 77, 136,
      140
  food security, 72, 75–78
    survey module, 50, 118, 163
Human subjects protection, 74
Hunger, 4, 5, 50, 53–55, 57–61, 70, 108, 110,
  111, 113–118, 180, 195, 201, 212, 221–225,
  235, 236
Hunger-Free Kids Act, 203
Hypoglycaemia treatment, 161
Hypothesis, 12, 14, 15, 48, 52, 70, 80, 81
Hypothesized mediator, 50, 52

I

ICF Macro Institutional Review Board, 49
Illicit drug, 106, 116
Immigrant communities, 179, 194
Impaired postnatal growth, 28
Implementation successes/challenges, 139,
  144
Independent variables, 13, 186
Indicators of low income and food insecurity,
  16, 20
Individual behavior, 203, 206, 211, 213, 227
Inductive-deductive approach, 140
Infant diarrhea, 28
Information gap, 37
Information-motivation-behavioral skills
  model, 92
Injection drug users, 106, 111, 113, 115
Integrated pest and disease management
  (IPM), 129
Inter-household commodity, 137
Intervention, 90–92, 94, 99, 100, 125,
  127–133, 136, 139–150, 206, 208, 210
Intervention model, 130
Intimate partner violence, 37, 38

IPM, 129, 130
Irrigation technology, 147

J

Jimma, 29, 30, 32, 36

K

Kaplan-Meir survival probabilities, 114
Kenyan Medical Research Institute, 132
KickStart, 128, 129
  Hip Pump, 129

L

La Via Campesina's members, 226
Latin American and Caribbean Food Security
  Scale, 94
Living Standards Measurement Study, 137
Loan Program, 129
Longitudinal data, 117
Low albumin and phosphate levels, 116
Low- and middle-income countries (LMICs),
  232
Low food security, 4, 5, 7, 180, 183, 184, 186,
  187, 193, 195, 200
Low infant birth weight, 28
Low-income ethnic minority populations, 202

M

Macro- and micro- food system environments,
  206
Macroenvironmental sectors, 208
Macroenvironmental settings, 207
Malnutrition, 105, 109, 116, 130, 131, 133,
  199, 231
Manifestation of emotional distress, 40
Mann–Whitney U tests, 14, 15
Marginal food security, 3
  anxiety over food sufficiency, 3
  shortage of food in the house, 3
Maternal distress, 27, 28, 38
Maternal health, 70, 73, 77, 79
Meals on Wheels, 174
Measurement scale, 37
Measurements on food insecurity, 13
  covariates, 14

data analysis, 14
dependent variable, 13
independent variables, 13
    diet, 14
    psychological functioning/symptoms,
    13, 14
Mediating variables, 52, 140, 141
Mediation and effect modification analyses,
    58
Medical Outcomes Study HIV Health Survey,
    135
Mental
    distress, 27–32, 35–40
    health, 117, 135, 142, 143
    illness, 15, 28, 36
Mestizo, 96
Metabolism, 22
Methods, 12, 29, 49, 71, 90, 106, 128, 162,
    181
Mexican
    heritage, 179
    immigrant, 179, 180, 194, 195
    marigold, 130
    origin children, 180, 181, 187, 193–195
    origin households, 179, 180, 193
    origin youth, 180
Microcredit
    loan, 129, 140
    programs, 127
Microenvironmental setting, 207, 208
Microfinance, 60, 127, 128, 129, 144, 145,
    148–150
Micronutrients, 12, 14
    folate, 12
    iron, 12
    vitamin B12, 12
    zinc, 12
Microwaveable foods, 168
Mid upper arm circumference measurements,
    134, 135
Ministry of Health and Child Care's
    (MoHCC), 71
Modeling, 32, 35, 51, 184, 212
Monetary incentives, 210

MoneyMaker Hip Pump, 129
Monitoring, 139, 161, 212
Mood disorders, 12, 15, 17, 20, 21, 23
Mortality, 99, 100, 105–107, 109, 110, 113,
    115–118, 125–127, 132, 137, 200
Mother-to-child transmission (MTCT),
    70–73, 78–83
Multisectoral intervention, 128, 131, 149
Multivariable regression analyses, 97
Multivariate logistic regression modeling
    analysis, 35
Multivariate models, 117

N

National Institutes of Health, 204
National School Lunch Program, 202, 203
    School Breakfast Programs, 194
Nausea, 92, 93
Non-communicable disease, 224, 233
Non-governmental organization, 128
Non-immigrant families, 180
Non-nucleoside reverse transcriptase inhibitor
    (NNRTI), 107
Non-parametric test, 185
Norwegian research, 175
Nucleoside reverse transcriptase inhibitor
    (NRTI), 107
Nutrient intakes, 12–14, 17–19, 21–23, 181,
    185, 187, 192, 193
Nutrient-poor energy-dense foods, 199
Nutrition Data System for Research, 185
Nutrition education curriculum, 92, 94
Nutrition gap, 8
Nutrition/weight status objectives, 203
Nutritional
    counseling, 90–95, 98–101
    knowledge, 90, 92, 94–99
    pathway, 131
    status, 89, 90, 92, 95, 99, 100, 106, 109,
    118, 135
NVivo data management software, 166

O

Obesity, 100, 162, 179, 180, 194, 199–213,
    222, 225, 231, 233–236

Obesity research task force, 204
Open data kit (ODK) collect, 139
Oral ulcers, 92
Overview of training provided to peer nutrition counselors, 94

P

Participant recruitment, 163
Partner violence, 36, 38
Patient-centered approach, 175
Pearson's Chi-Square tests, 110
Peasant movements, 225
PepsiCo, 226, 227, 231, 234
Pesquisa Nacional de Demografia e Saúde da Criança e da Mulher (PNDS), 49
Pesticide companies, 225
Philanthropic foundations, 226
Phlebotomy, 134
Physical food environment, 208, 209
PLHIV, 90, 99–101, 125–127, 130, 132, 150
PLoS Medicine series, 231
PMTCT cascade, 69–73, 76–79, 81, 82
Poisson
    model, 15, 17
    regression, 15, 17, 73
Policy
    gaps, 213
    level changes, 203
    makers, 11, 39, 199, 202
Political food environment, 211
Polychoric correlation matrix, 74
Pooled logistic regression, 141
Poor dietary diversity, 106
Poorer cognitive development, 28
Population Survey Food Security Supplement, 6, 180
Portable stadiometer, 183
Potassium, 180, 185, 186
Poverty, 14, 38, 80, 125, 127, 131, 162, 179, 182, 194, 201, 224, 231, 236
Practitioners, 11, 199
Pre- and post-intervention changes in food insecurity, 98
Pregnancy, 27, 28, 37, 38, 40, 71, 72, 74, 76, 79, 81, 82

Pregnant women, 28–30, 32, 35–40, 69, 70, 79, 83
Prenatal depressive symptoms, 28
Prevalence of household food insecurity/ mental distress, 35
Prevalence ratios (PR), 73
Prevention of mother-to-child HIV transmission, 69
Primary explanatory variable, 50, 107
Primary health outcomes, 141
Private–public partnerships, 235
Probability proportional to size sampling (PPS) technique, 29
Process evaluation methods, 139, 149
Professional nutritionists, 91, 92, 101
Promotora-researchers, 182, 183
Protease inhibitor, 107, 111
Protein, 17, 21, 82, 93, 95, 135, 148, 172, 185, 186
Psychiatric
    emergency unit, 11
    symptoms, 22
        anxiety, 22
        depression, 22
        suicide ideation, 22
Psychological
    function, 12, 21, 22
    issues, 11
Public health
    goals and research priorities, 203
    nutrition goals, 201, 206
    organizations, 118
    success, 204, 213
Pump challenges, 147

Q

Qualitative methodology, 162
Qualitative software, 140
Questionnaire to measure household food insecurity access scale, 31

R

Radimer/Cornell scale, 108, 116–118
Randomization, 91, 92, 99, 132

Randomized controlled trial, 91, 128, 132, 148
Regression methods, 141
Reguladores, 95
Resource-intensive farming, 226
Risk populations, 200, 205, 209, 211, 213
Ritonavir, 107
Role of markets and governments, 226

S

School-based programs, 211
Screening and enrollment, 141, 148
Secondary explanatory variables, 109
Seed selection, 129
Self-care behaviours, 175
   blood glucose monitoring, 175
   healthy coping, 175
   physical activity, 175
Self-efficacy, 60, 175
Self-reporting questionnaire, 30
Self-sufficiency, 210, 223
Sexual relationship power scale, 136
Sexual risk behavior, 142
Shamba Maisha, 125, 127, 128, 131, 132, 134, 139, 142, 144–149
SNAP retailers, 209
Social desirability bias, 118
Social Determinants of Health Framework, 162
Social isolation, 167, 170, 174
Social support, 38, 126, 134, 138, 162, 175
Socio demographic characteristics, 32
Sociocultural food environment, 211
Sociodemographic
   characteristics of participants, 75, 95, 163
   variables, 142
Socioeconomic
   determinants, 105
   household wealth, 33
   status, 99, 132, 149
Sodium foods, 168
Soil and water conservation, 129
Sovereignty, 223, 224
Stand-alone strategy, 127
Standard errors, 185

Stata Statistical Software, 185
Statistical
   analysis, 50, 110
   methods, 140
   tests, 140
Status quo gap, 8
Stunted growth, 28
Sub-Saharan Africa, 7, 38, 47, 69, 125, 126, 131, 135, 149
Successes with Shamba Maisha implementation, 144
Sugar sweetened beverages (SSBs), 233
Summary score, 143
Supplemental Nutrition Assistance Program, 202
Survey and anthropometric measurements, 93
Sweeteners, 168, 173, 180
Systematic approach, 204, 206
Systemic inequity, 225

T

Task Force, 203
Telescoping errors, 81
Temporary Assistance for Needy Families Program, 210
Texas A&M University Institutional Review Board, 182
Texas border region, 180, 194
Theoretical model, 127
Therapeutic diet, 14
Third National Health and Nutrition Examination Survey, 60
Tools and measurements, 30
T-tests, 140
Type 2 diabetes, 174, 175, 180, 194, 233

U

Under nutrition, 28
Undernourishment, 221, 222, 224
United States Department of Agriculture, 200
Univariate analysis, 116
USDA'S labels on food security, 3, 4
   food insecurity, 4
      low food security, 4
      very low food security, 4

food security, 3
    anxiety over food sufficiency , 3
    high food security, 3
    marginal food security, 3
    shortage of food, 3

V

Variable selection, 107
    primary explanatory variable, 107, 108
    secondary explanatory variables, 109
Very low food security, 5, 183, 186, 193, 195, 200
Viral load suppression, 141, 149
Viral suppression, 126
Virologic non-suppression, 106
Vitamin C, 17, 21, 22, 186
Vitamin D, 180, 185, 186, 187
Vulnerable populations, 135

W

Wald test, 76

Wald-type F-tests, 52
White House Task Force on Childhood Obesity, 203
Wilcoxon Rank Sum Test, 110, 140
Wilcoxon Signed-Rank Test, 185, 191
Women's disempowerment, 224
Women's empowerment, 127, 131, 134
Wood ash, 130
World Food Programme, 90, 126, 135
World Food Summit, 222
World Health Organization, 30, 69, 89, 95, 126
World Trade Organization, 226

Y

Young Mania rating scale, 13, 20

Z

Zimbabwe, 69, 70, 71, 74, 75, 77–80, 82, 83
Zinc, 12, 14, 17, 21, 116

T - #0845 - 101024 - C280 - 229/152/12 - PB - 9781774636886 - Gloss Lamination